MACMILLAN/McGRAW-HILL

Math

 **Macmillan
McGraw-Hill**

PROGRAM AUTHORS

Douglas H. Clements, Ph.D.

Professor of Mathematics Education

State University of
New York at Buffalo

Buffalo, New York

Carol E. Malloy, Ph.D.

Assistant Professor of
Mathematics Education

University of North Carolina at
Chapel Hill

Chapel Hill, North Carolina

Lois Gordon Moseley

Mathematics Consultant

Houston, Texas

Yuria Orihuela

District Math Supervisor

Miami-Dade County Public Schools

Miami, Florida

Robyn R. Silbey

Montgomery County Public Schools

Rockville, Maryland

SENIOR CONTENT REVIEWERS

Gunnar Carlsson, Ph.D.

Professor of Mathematics

Stanford University

Stanford, California

Ralph L. Cohen, Ph.D.

Professor of Mathematics

Stanford University

Stanford, California

The McGraw·Hill Companies

 **Macmillan
McGraw-Hill**

Published by Macmillan/McGraw-Hill, of McGraw-Hill Education, a division of The McGraw-Hill Companies, Inc., Two Penn Plaza, New York, New York 10121.

Copyright © 2005 by Macmillan/McGraw-Hill. All rights reserved. No part of this publication may be reproduced or distributed in any form or by any means, or stored in a database or retrieval system, without the prior written consent of The McGraw-Hill Companies, Inc., including, but not limited to, network storage or transmission, or broadcast for distance learning.

Foldables™, Math Tool Chest™, Math Traveler™, Mathematics Yes!™, Yearly Progress Pro™, and Math Background for Professional Development™ are trademarks of The McGraw-Hill Companies, Inc.

Printed in the United States of America

ISBN 0-02-105011-2
 4 5 6 7 8 9 073.08 07 06 05

learning through listening

Students with print disabilities may be eligible to obtain an accessible, audio version of the pupil edition of this textbook. Please call Recording for the Blind & Dyslexic at 1-800-221-4792 for complete information.

READ TOGETHER

BUSY BUGS INDEED!

Story by Jenny Nichols
Illustrated by Elise Mills

Look at the bugs working hard.
They all want to clean the yard.

How many carry green leaves?

How many carry red leaves?

All the bugs carry leaves.
They are busy bugs, indeed!

How many bugs carry leaves? ☐

© Macmillan/McGraw-Hill

Lots of worms pull carts with grass.
They are smiling as they pass.
2 carts of hay move very slow.
3 more of dirt are set to go.

How many worms
pull carts of dirt and hay?

Worms pulling carts with grass smile as they pass.

6 bugs have saws and 4 have leaves. These are busy bugs, indeed.

How many busy bugs do the worms pass?

Math at Home

Dear Family,

I will learn ways to add sums to 18 in Chapter 17. Here are my math words and an activity that we can do together.

Love, _____

My Math Words

doubles :

$$\begin{array}{r} 7 \\ +7 \\ \hline 14 \end{array}$$

doubles plus 1 :

$$\begin{array}{r} 7 \\ +8 \\ \hline 15 \end{array}$$

addends :

$$7 + 4 + 5 = 16$$
$\uparrow \quad \uparrow \quad \uparrow$ **addends**

sum :

$$9 + 8 = 17 \longleftarrow \text{sum}$$

Home Activity

Write these numbers on cards.

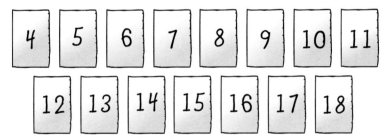

Your child picks a card and tells what number is 1 more and 1 less than that number. Then have your child tell what number is 10 more and 10 less.

© Macmillan/McGraw-Hill

Books to Read

Look for these books at your local library and use them to help your child learn addition facts.

- **1 Is One** by Tasha Tudor, Simon & Schuster, 1984.
- **Ten Friends** by Bruce Goldstone, Henry Holt and Company, 2001.

www.mmhmath.com
For Real World Math Activities

Learn I know many facts with sums to 12.

I can count on 1, 2, or 3.

I can show related addition facts.

I can add 0.

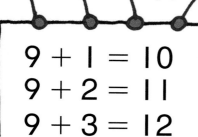

$9 + 1 = 10$
$9 + 2 = 11$
$9 + 3 = 12$

$8 + 3 = 11$
$3 + 8 = 11$

$9 + 0 = 9$

Try It Find each sum.

1. $3 + 9 = \underline{12}$ 2. $6 + 4 = \underline{}$ 3. $2 + 7 = \underline{}$

4. $7 + 0 = \underline{}$ 5. $7 + 1 = \underline{}$ 6. $4 + 6 = \underline{}$

7. $0 + 8 = \underline{}$ 8. $7 + 2 = \underline{}$ 9. $9 + 3 = \underline{}$

10. $\begin{array}{r} 9 \\ +1 \\ \hline \end{array}$ 11. $\begin{array}{r} 5 \\ +6 \\ \hline \end{array}$ 12. $\begin{array}{r} 6 \\ +2 \\ \hline \end{array}$ 13. $\begin{array}{r} 0 \\ +7 \\ \hline \end{array}$ 14. $\begin{array}{r} 8 \\ +3 \\ \hline \end{array}$ 15. $\begin{array}{r} 3 \\ +7 \\ \hline \end{array}$

16. **Write About It!** Add $9 + 2$. Tell how you found the sum.

© Macmillan/McGraw-Hill

Practice Add. Then color.

Sums of 7 or 8 ▸ Sums of 9 or 10 ▸ Sums of 11 or 12 ▸

7 + 4 = ___

3 + 8 = ___

9 + 1 = ___

2 + 8 = ___

$$\begin{array}{r} 4 \\ +3 \\ \hline \end{array}$$

$$\begin{array}{r} 6 \\ +1 \\ \hline \end{array}$$

1 + 6 = ___

3 + 4 = ___

6 + 5 = ___

4 + 7 = ___

$$\begin{array}{r} 6 \\ +4 \\ \hline \end{array}$$

$$\begin{array}{r} 3 \\ +7 \\ \hline \end{array}$$

2 + 6 = ___

7 + 1 = ___

Math at Home: Your child practiced addition facts to 12.
Activity: Have your child write five facts with a sum of 12.

Name_____

Learn These pictures show doubles.

Math Words

doubles

addend

The addends are the same.

   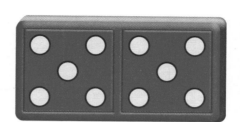

6 + 6 = 12   5 + 5 = 10

Try It Draw the missing dots to show a double.
Then write the doubles fact.

1.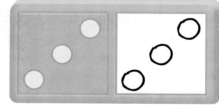

____3____ + ___3___ = ___6___

2.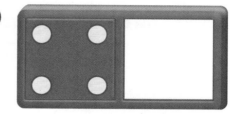

____ + ____ = ____

3.

____ + ____ = ____

4.

____ + ____ = ____

5. ✏️ Write **About It!** What doubles can you show with
your hands?

© Macmillan/McGraw-Hill

Draw dots to show the doubles.
Write the addends.

Look at the sum to find the doubles fact.

6
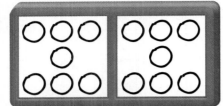

_7_ + _7_ = 14

7

___ + ___ = 12

8

___ + ___ = 10

9

___ + ___ = 16

10 4
+4

11 9
+9

12 5
+5

13 7
+7

14 6
+6

15 8
+8

Problem Solving — Number Sense

16 I have 8 cars.
Tim has 8 cars.
How many cars do we
have in all?

____ + ____ = ____ cars

17 Jack has 7 cars.
Ken has the same
number of cars.
How many cars do they
have in all?

____ + ____ = ____ cars

Math at Home: Your child added doubles to 18.
Activity: Have your child find things around the house that show doubles, such
as eggs in a carton (6 + 6 = 12), juice boxes (3 + 3 = 6), or toes on two feet (5 + 5 = 10).

Name_____

Doubles Plus 1

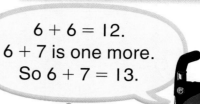

Learn Find each sum.
Use doubles to help you add.

Math Word
doubles plus 1

6 + 6 = 12.
6 + 7 is one more.
So 6 + 7 = 13.

6 + 6 = **12**   6 + 7 = **13**

doubles   doubles plus 1

Your Turn Find each sum. Use .

1

7 + 7 = **14**   7 + 8 = ____

2

5 + 5 = ____   5 + 6 = ____

3 ✏️ **Write About It!** How does knowing 8 + 8 help you find 8 + 9?

© Macmillan/McGraw-Hill

Circle the doubles.
Find each sum. Use 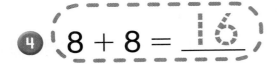.

Doubles can help you add.

4. (8 + 8 = __16__)

8 + 9 = __17__

9 + 8 = __17__

5. 3 + 3 = ____   3 + 4 = ____   4 + 3 = ____

6. 2 + 2 = ____   2 + 3 = ____   3 + 2 = ____

7.
```
  7       7       8
 +7      +8      +7
```
8.
```
  4       4       5
 +4      +5      +4
```

Problem Solving   **Number Sense**

 THINK SOLVE EXPLAIN

Write a doubles plus 1 fact to solve.
What doubles fact can help you?

9. Jan sees 4 🕷.
   Tim sees 5 🕷.
   How many 🕷 do they see in all?

4 + 5 = ____ 🕷   ____ + ____ = 8 🕷

 Math at Home: Your child used doubles to learn doubles plus one facts.
Activity: Show two stacks of six dimes each. Ask your child to tell the doubles fact.
Add one dime and have your child tell the new addition fact.

Name_____

Add Three Numbers

Learn You can use doubles or make a ten to add three numbers.

Add the doubles first.

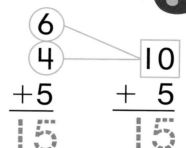

Add to make a ten first.

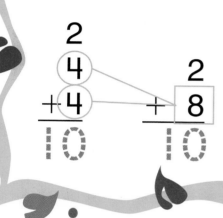

$$\begin{array}{r} 2 \\ 4 \\ +4 \\ \hline 10 \end{array} \qquad \begin{array}{r} 2 \\ +8 \\ \hline 10 \end{array}$$

$$\begin{array}{r} 6 \\ 4 \\ +5 \\ \hline 15 \end{array} \qquad \begin{array}{r} 10 \\ + 5 \\ \hline 15 \end{array}$$

Try It Add the doubles or make a ten first. Then find the sum.

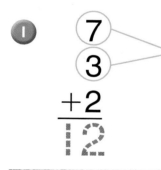

1.
$$\begin{array}{r} 7 \\ 3 \\ +2 \\ \hline 12 \end{array} \qquad 10$$

2.
$$\begin{array}{r} 7 \\ 7 \\ +1 \\ \hline \end{array} \qquad \square$$

3.
$$\begin{array}{r} 5 \\ 1 \\ +9 \\ \hline \end{array} \qquad \square$$

4.
$$\begin{array}{r} 8 \\ 8 \\ +1 \\ \hline \end{array} \qquad \square$$

5.
$$\begin{array}{r} 3 \\ 2 \\ +8 \\ \hline \end{array} \qquad \square$$

6.
$$\begin{array}{r} 9 \\ 9 \\ +0 \\ \hline \end{array} \qquad \square$$

© Macmillan/McGraw-Hill

7. ✎ Write **About It!** What are two ways to add $8 + 2 + 8$?

Practice

Circle two numbers to add first. Then find the sums.

(3)
5
$+(3)$ ➤ (6)
11

$+5$
11

Look for doubles.

(3)
2
$+(7)$ ➤ (10)
12

$+2$
12

Look to make a ten.

---

8 (7)
(7)
$+2$
16

9 6
4
$+5$

10 4
4
$+9$

11 3
7
$+8$

12 2
8
$+4$

---

13 2
6
$+6$

14 5
7
$+5$

15 6
6
$+1$

16 4
8
$+6$

17 7
8
$+2$

---

18 7
7
$+1$

19 9
7
$+1$

20 6
5
$+5$

21 8
9
$+1$

22 1
9
$+3$

Problem Solving (Critical Thinking

Choose the best strategy. Circle it. Solve.

23 $4 + 6 + 3 =$ _____   make a ten   doubles

Math at Home: Your child learned to add three numbers by adding doubles and by making a ten.
Activity: Write the numbers 6, 6, and 2 on a sheet of paper. Ask your children to circle the numbers they would add first and then find the sum. Do the same for the numbers 8, 6, and 2.

Name_____

Add Three Numbers in Any Order

ALGEBRA

Learn You can add three numbers in any order.
The sum stays the same.

Add 9 + 1.
Then add 2.

$$\begin{array}{r} ⑨ \\ ① \\ +2 \\ \hline 12 \end{array}$$ 10

Add 9 + 2.
Then add 1.

$$\begin{array}{r} ⑨ \\ 1 \\ +② \\ \hline 12 \end{array}$$ 11

Add 1 + 2.
Then add 9.

$$\begin{array}{r} 9 \\ ① \\ +② \\ \hline 12 \end{array}$$ 3

Try It Circle two numbers to add first.
Then find the sums.

1.

$$\begin{array}{r} 7 \\ 1 \\ +2 \\ \hline 10 \end{array}$$ 8

$$\begin{array}{r} 7 \\ 1 \\ +2 \\ \end{array}$$ 9

$$\begin{array}{r} 7 \\ 1 \\ +2 \\ \end{array}$$ 3

2.

$$\begin{array}{r} 3 \\ 6 \\ +2 \\ \end{array}$$ ☐

$$\begin{array}{r} 3 \\ 6 \\ +2 \\ \end{array}$$ ☐

$$\begin{array}{r} 3 \\ 6 \\ +2 \\ \end{array}$$ ☐

3. ✎ Write **About It!** Why would you get the same sum for
5 + 2 + 1 and 1 + 2 + 5?

Chapter 17 Lesson 5

© Macmillan/McGraw-Hill

three hundred three **303**

```
 3      5      2
 5      3      3
+2     +2     +5
─────  ─────  ─────
 10     10     10
```

I can add 3, 5, and 2 in any order. The sum is the same.

4
```
 9      9      8
 0      8      0
+8     +0     +9
─────  ─────  ─────
 17     17     17
```

5
```
 2      2      8
 1      8      2
+8     +1     +1
─────  ─────  ─────
```

6
```
 3      4      1
 1      1      3
+4     +3     +4
─────  ─────  ─────
```

7
```
 3      5      1
 1      1      5
+5     +3     +3
─────  ─────  ─────
```

Problem Solving ⟩ Number Sense

8 Tony has 3 🪙 and 2 🪙.
He wants to buy 3 🚚.
Each 🚚 costs 10 ¢.
Does Tony have enough money to buy 3 🚚? Explain.

_____

_____

Math at Home: Your child learned to add numbers in any order.
Activity: Write the numbers 9, 8, and 1 on cards. Have your child add the numbers telling you the order he or she used to add. Then have your child add the same numbers in a different order.

Problem Solving Skill
Reading for Math

A Family Hike

Pat and her dad saw
many butterflies.
First, they saw 6 red ones.
Next, they saw 5 blue ones.

Then, a bright yellow butterfly
landed on Pat's arm!

Problem Solving

 **Sequence of Events**

1. What color butterfly did
   Pat and her dad see first? _____

2. What was the last thing that happened?

   _____

3. How many red and blue butterflies
   did Pat and her dad see? _____ butterflies

   Write a number sentence.

   ____ ◯ ____ ◯ ____

4. How many butterflies did they see in all? _____ butterflies
   Write a number sentence.

   ____ ◯ ____ ◯ ____ ◯ ____

© Macmillan/McGraw-Hill

So Many Ants!

First Pat saw 6 ants on a rock.
Next she saw 4 ants on a leaf.
Then 8 ants crawled on her leg.

Problem Solving

Reading Skill Sequence of Events

5 How many ants did Pat see first? _____ ants

6 How many ants did Pat see
on the rock and on the leaf? _____ ants

Write a number sentence.

_____ ◯ _____ ◯ _____

7 How many ants were on Pat's leg? _____ ants

8 How many ants did Pat see in all? _____ ants

Write a number sentence.

_____ ◯ _____ ◯ _____ ◯ _____

Math at Home: Your child followed a sequence of events to answer questions.
Activity: Ask your child to tell a story about seeing a group of seven and later seeing a group of six ladybugs.
Then have your child write a number sentence to tell how many in all.

Name_____

Problem Solving Practice

Solve.

---

1 There are 6 🦋 on a flower. 6 more 🦋 land. How many 🦋 are there in all? _____ 🦋

2 10 🐜 are on a rock. 5 🐜 are in the grass. How many 🐜 are there in all? _____ 🐜

---

3 Ben sees 9 red butterflies. Then he sees 2 blue butterflies. How many butterflies does he see in all?

_____ + _____ = _____

_____ butterflies

4 7 ladybugs are in the garden. 8 ladybugs are on the grass. How many ladybugs are there in all?

_____ + _____ = _____

_____ ladybugs

Problem Solving

 Write a Story!

5 Find the sum.

Write an addition story about the number sentence.

$8 + 9 =$ _____

_____

_____

_____

© Macmillan/McGraw-Hill

Writing for Math

THINK
SOLVE
EXPLAIN

Max sees 6 bugs on each green leaf.
He sees 3 bugs on the yellow leaf.
How many bugs does he
see in all?

Writing

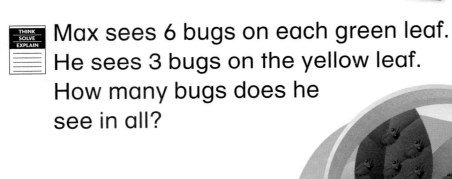

Think

What numbers would I use to solve the problem?

_____ 🍃 _____ 🍃 _____ 🍃

Solve

How can I use the numbers to find how many in all?

_____

____ ◯ ____ ◯ ____ = ____

Explain

I can tell you how my number sentence solves the problem.

_____

e-Journal **www.mmhmath.com**
Write about math

Name_____

Add.

1   8
     +8

2   8
     +9

3   7
     +7

4   7
     +8

5   6
     +5

6   9
     +8

7   8
     +7

8   7
     +6

9   6
     +5

10   3
     2
    +8

11   6
     2
    +4

12   9
     0
    +9

13   7
     1
    +7

14   2
     1
    +9

15   2
     6
    +6

16   5
     5
    +7

17   3
     7
    +8

18   4
     4
    +9

19   8
     2
    +7

20 First, Brittany had 4 pens.
Next, her mom gave her 10 more.
How many pens does Brittany have now? _____ pens

© Macmillan/McGraw-Hill

Assessment

Spiral Review and Test Prep
Chapters 1-17

Choose the best answer.

1 Count the money.

    3¢   26¢   36¢

    ◯   ◯   ◯

2 Skip-count by 2s. What are the missing numbers?

62	64	66			72

    67, 68   74, 76   68, 70

    ◯   ◯   ◯

3 Molly eats 6 . Adam eats 7 .
How many do they eat in all?

    12   13   14

    ◯   ◯   ◯

Solve.

4 $8 + 4 = $ _____   **5** $11 - 6 = $ _____

6 Tyrone has 40¢.

Show another way to make 40¢.

Test Prep

Subtraction Strategies and Facts to 18

SING TOGETHER

Butterfly Song

Sung to the tune of "Rock-a-Bye Baby"

When I see butterflies

Flying so low,

I try to count them

Before they go.

Eighteen touch down,

Then nine fly away—

Only nine left

To brighten our day!

Math at Home

Dear Family,

I will learn ways to subtract from numbers through 18 in Chapter 18. Here are my math words and an activity that we can do together.

Love, _____

My Math Words

related facts :

$8 + 8 = 16$ $16 - 8 = 8$

related subtraction facts :

$15 - 7 = 8$ $15 - 8 = 7$

fact family :

$5 + 9 = 14$ $14 - 9 = 5$
$9 + 5 = 14$ $14 - 5 = 9$

Home Activity

Show these cards:

| $6 + \underline{} = 10$ | $4 + \underline{} = 9$ | $8 + \underline{} = 12$ |

| $7 + \underline{} = 11$ | $6 + \underline{} = 12$ |

Have your child find the missing addend. Then have your child use any small household items to show how he or she solved the problems.

© Macmillan/McGraw-Hill

Books to Read

Look for these books at your local library and use them to help your child subtract from 18.

- **Math at the Store** by William Amato, Children's Press, 2002.
- **Bunny Money** by Rosemary Wells, Viking, 1997.

LOG ON
www.mmhmath.com
For Real World Math Activities

Practice Subtraction Facts

Learn I know many ways to subtract from numbers through 12.

I can count back 1, 2, or 3.

I can use related subtraction facts.

I can subtract 0 or subtract all.

$10 - 1 = 9$
$10 - 2 = 8$
$10 - 3 = 7$

$11 - 3 = 8$
$11 - 8 = 3$

$9 - 0 = 9$
$9 - 9 = 0$

Try It Find each difference.

1. $9 - 3 = \underline{6}$
2. $12 - 9 = \underline{}$
3. $7 - 2 = \underline{}$

4. $12 - 3 = \underline{}$
5. $8 - 1 = \underline{}$
6. $11 - 2 = \underline{}$

7. $11 - 4 = \underline{}$
8. $8 - 8 = \underline{}$
9. $11 - 7 = \underline{}$

10. $\begin{array}{r} 12 \\ -5 \\ \hline \end{array}$
11. $\begin{array}{r} 8 \\ -0 \\ \hline \end{array}$
12. $\begin{array}{r} 9 \\ -2 \\ \hline \end{array}$
13. $\begin{array}{r} 12 \\ -4 \\ \hline \end{array}$
14. $\begin{array}{r} 9 \\ -1 \\ \hline \end{array}$
15. $\begin{array}{r} 7 \\ -0 \\ \hline \end{array}$

16. Write **About It!** Subtract $12 - 7$. What related subtraction fact do you know?

© Macmillan/McGraw-Hill

Practice Subtract. Then color.

Differences 0 to 5  Differences 6 to 9

$11 - 7 =$ _____

$5 - 0 =$ _____

$12 - 7 =$ _____

$$\begin{array}{r} 11 \\ -\ 5 \\ \hline \end{array}$$

$12 - 4 =$ _____

$$\begin{array}{r} 12 \\ -\ 4 \\ \hline \end{array}$$

$$\begin{array}{r} 8 \\ -\ 0 \\ \hline \end{array}$$

$$\begin{array}{r} 10 \\ -\ 3 \\ \hline \end{array}$$

$$\begin{array}{r} 9 \\ -\ 2 \\ \hline \end{array}$$

$$\begin{array}{r} 9 \\ -\ 6 \\ \hline \end{array}$$

$$\begin{array}{r} 6 \\ -\ 5 \\ \hline \end{array}$$

$$\begin{array}{r} 8 \\ -\ 6 \\ \hline \end{array}$$

$$\begin{array}{r} 12 \\ -\ 8 \\ \hline \end{array}$$

$$\begin{array}{r} 10 \\ -\ 7 \\ \hline \end{array}$$

$$\begin{array}{r} 8 \\ -\ 5 \\ \hline \end{array}$$

Math at Home: Your child practiced subtraction facts.
Activity: Write ten examples from this lesson on a sheet of paper.
Have your child tell you how to solve each problem.

Name_____

 **Learn** You can use addition doubles to help you subtract.

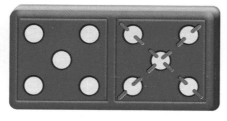 I know 5 + 5 = 10.
So 10 − 5 = 5.

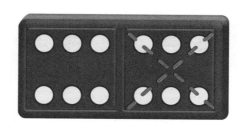 I know 6 + 6 = 12.
So 12 − 6 = 6.

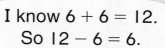

Try It Add the double. Then subtract.

① 3 + 3 = __6__

6 − 3 = __3__

② 8 + 8 = _____

16 − 8 = _____

③ 5 + 5 = _____

10 − 5 = _____

④ 9 + 9 = _____

18 − 9 = _____

⑤ 7 + 7 = _____

14 − 7 = _____

⑥ 4 + 4 = _____

8 − 4 = _____

⑦ Write **About It!** What addition double can help you
find 8¢ − 4¢?

© Macmillan/McGraw-Hill

Add or subtract. Then draw
a line to match the related facts.

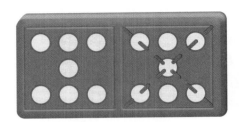 $7 + 7 = \underline{14}$

$14 - 7 = \underline{7}$

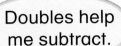 Doubles help
me subtract.

8 $6 + 6 = \underline{12}$ ---------- $18 - 9 = \underline{}$

9 $2 + 2 = \underline{}$ ---------- $12 - 6 = \underline{6}$

10 $9 + 9 = \underline{}$ $8 - 4 = \underline{}$

11 $4 + 4 = \underline{}$ $4 - 2 = \underline{}$

12 $8 + 8 = \underline{}$ $16 - 8 = \underline{}$

x **Algebra • Patterns**

13 Draw the missing dots to make a double.
Then write the related fact.

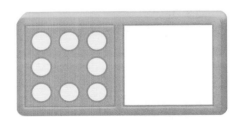

$\underline{} + \underline{} = \underline{}$

$\underline{} - \underline{} = \underline{}$

Math at Home: Your child used doubles to subtract.
Activity: Write several subtraction facts such as 12 − 6 = _____, on separate cards. Ask your child to write the doubles fact that helps solve the problem. Have your child use stacks of pennies to show the facts.

Related Subtraction Facts

Learn You can use related subtraction facts to help you subtract.

Math Word
_____
related subtraction facts

Related facts use the same numbers.

$13 - 6 = \underline{7}$ $13 - 7 = \underline{6}$

Try It Complete the related subtraction facts for each picture.

1.

$15 - 7 = \underline{8}$

$15 - 8 = \underline{7}$

2.

$17 - 8 = \underline{\hphantom{0}}$

$17 - 9 = \underline{\hphantom{0}}$

3.

$11 - 5 = \underline{\hphantom{0}}$

$11 - 6 = \underline{\hphantom{0}}$

4. Write **About It!** $12 - 7 = 5$. What is the related subtraction fact?

© Macmillan/McGraw-Hill

Practice Complete the related subtraction facts for each picture.

Related facts use the same numbers.

5

$11 - 4 = \underline{7}$

$11 - 7 = \underline{4}$

6

$12 - 8 = \underline{\hphantom{0}}$

$12 - 4 = \underline{\hphantom{0}}$

Subtract.

7 $10 - 6 = \underline{\hphantom{0}}$

$10 - 4 = \underline{\hphantom{0}}$

8 $17 - 9 = \underline{\hphantom{0}}$

$17 - 8 = \underline{\hphantom{0}}$

9 $13 - 7 = \underline{\hphantom{0}}$

$13 - 6 = \underline{\hphantom{0}}$

10 $9 - 7 = \underline{\hphantom{0}}$

$9 - 2 = \underline{\hphantom{0}}$

11 $11 - 5 = \underline{\hphantom{0}}$

$11 - 6 = \underline{\hphantom{0}}$

12 $15 - 7 = \underline{\hphantom{0}}$

$15 - 8 = \underline{\hphantom{0}}$

x Algebra • Missing Numbers

13 Use the picture to find the missing number.

$10 - \boxed{} = 9 \qquad 10 - \boxed{} = 1$

Math at Home: Your child learned about related subtraction facts.
Activity: Have your child use pennies to show 10 − 8 = 2. Then have him or her show
the related subtraction fact. Repeat with 12 − 9 = 3.

Name_____

 **Relate Addition and Subtraction**

ALGEBRA

Learn Related facts use the same numbers.

First, subtract one color. Then subtract the other color.

$4 + 9 = 13$

$13 - 4 = 9$ $13 - 9 = 4$

This cube train shows the numbers 4, 9, and 13.

Try It Use  and . Add. Then subtract. Write the related subtraction facts.

Add.	Subtract.	Write the related subtraction facts.
1 $5 + 8 = \underline{13}$	8	$13 \bigcirc 8 \bigcirc 5$
	5	___ $\bigcirc$ ___ $\bigcirc$ ___
2 $6 + 9 = \underline{}$	9	___ $\bigcirc$ ___ $\bigcirc$ ___
	6	___ $\bigcirc$ ___ $\bigcirc$ ___

3 ✏️ Write **About It!** If you know $8 + 6 = 14$, what related subtraction facts do you know?

 © Macmillan/McGraw-Hill

Practice Use cubes. Add. Then subtract. Write the related subtraction facts.

> Related facts use the same numbers.

Add.	Subtract.	Write the related subtraction facts.
4 $8 + 9 = \underline{17}$	9	$\underline{17} \ominus \underline{9} \ominus \underline{8}$
	8	$\underline{\quad} \bigcirc \underline{\quad} \bigcirc \underline{\quad}$

5 $5 + 9 = \underline{14}$

$\underline{14} \ominus \underline{9} \ominus \underline{5}$

$\underline{14} \ominus \underline{5} \ominus \underline{9}$

6 $7 + 8 = \underline{\quad}$

$\underline{\quad} \bigcirc \underline{\quad} \bigcirc \underline{\quad}$

$\underline{\quad} \bigcirc \underline{\quad} \bigcirc \underline{\quad}$

7 $8 + 6 = \underline{\quad}$

$\underline{\quad} \bigcirc \underline{\quad} \bigcirc \underline{\quad}$

$\underline{\quad} \bigcirc \underline{\quad} \bigcirc \underline{\quad}$

8 $9 + 7 = \underline{\quad}$

$\underline{\quad} \bigcirc \underline{\quad} \bigcirc \underline{\quad}$

$\underline{\quad} \bigcirc \underline{\quad} \bigcirc \underline{\quad}$

✓ Spiral Review and Test Prep

Choose the best answer.

9 30 is _____ than 37

greater ⬭   less ⬭

10 $8 + \underline{\quad} = 12$

4 ⬭   8 ⬭

Math at Home: Your child learned how addition and subtraction are related.
Activity: Have your child use small household objects, such as beans or buttons, to show how 5 + 8 = 13 is related to 13 − 8 = 5 and 13 − 5 = 8.

Name_____

Learn A fact family uses the same numbers.

Math Word

fact family

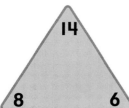

$8 + 6 = \underline{14}$ $14 - 6 = \underline{8}$

$6 + 8 = \underline{14}$ $14 - 8 = \underline{6}$

6, 8, and 14 make up this fact family.

Try It Add and subtract.
Complete each fact family.

1.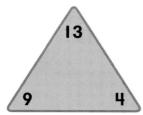

$8 + 7 = \underline{15}$ $15 - 7 = \underline{8}$

$7 + 8 = \underline{15}$ $15 - 8 = \underline{7}$

2.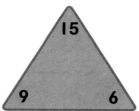

$9 + 6 = \underline{}$ $15 - 6 = \underline{}$

$6 + 9 = \underline{}$ $15 - 9 = \underline{}$

3.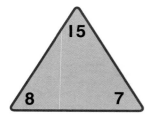

$9 + 4 = \underline{}$ $13 - 4 = \underline{}$

$4 + 9 = \underline{}$ $13 - 9 = \underline{}$

4.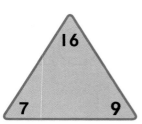

$7 + 9 = \underline{}$ $16 - 9 = \underline{}$

$9 + 7 = \underline{}$ $16 - 7 = \underline{}$

5. Write **About It!** What fact family can you make with the numbers 8, 8, and 16?

© Macmillan/McGraw-Hill

Practice

Add and subtract.
Complete each fact family.

A fact family uses the same numbers.

6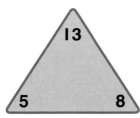

$5 + 8 = \underline{13}$ $13 - 8 = \underline{5}$

$8 + 5 = \underline{\hphantom{00}}$ $13 - 5 = \underline{\hphantom{00}}$

7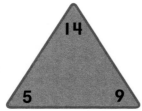

$5 + 9 = \underline{\hphantom{00}}$ $14 - 9 = \underline{\hphantom{00}}$

$9 + 5 = \underline{\hphantom{00}}$ $14 - 5 = \underline{\hphantom{00}}$

8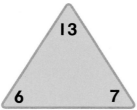

$6 + 7 = \underline{\hphantom{00}}$ $13 - 7 = \underline{\hphantom{00}}$

$7 + 6 = \underline{\hphantom{00}}$ $13 - 6 = \underline{\hphantom{00}}$

9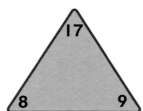

$8 + 9 = \underline{\hphantom{00}}$ $17 - 9 = \underline{\hphantom{00}}$

$9 + 8 = \underline{\hphantom{00}}$ $17 - 8 = \underline{\hphantom{00}}$

Problem Solving   Critical Thinking

10 Write the fact family that tells about the picture. Share your work with a friend.

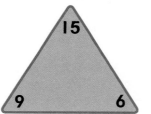

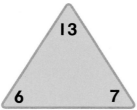

Math at Home: Your child learned about fact families.
Activity: Have your child draw a row of 8 red flowers and a row of 2 blue flowers. Then have the child write the fact family that matches the picture.

322   three hundred twenty-two

Problem Solving Strategy

Name_____

Choose the Operation • Algebra

You can choose the operation to solve a problem.

Kim sees 17 balloons.
Ben sees 9 balloons.
How many more balloons does Kim see?

Read

What do I already know?   Kim sees _____ balloons.

Ben sees _____ balloons.

What do I need to find?

_____

Plan

Do I add or subtract to solve the problem?   + add   − subtract

Solve

I can carry out my plan. _____ _____ _____ balloons

Kim sees _____ more balloons than Ben.

Look Back

Does my answer make sense?   Yes.   No.

How do I know? _____

Problem Solving

© Macmillan/McGraw-Hill

Circle add or subtract.
Write a number sentence to solve.

Draw or write to explain.

1 14 bees are in the garden.
8 bees fly away.
How many bees are left?

+		−
add		subtract

_____ ◯ _____ ◯ _____

bee

---

2 9 birds are in a tree.
4 birds are in a bush.
How many birds are
there in all?

+		−
add		subtract

_____ ◯ _____ ◯ _____

bird

---

3 15 ladybugs are by a rock.
9 ladybugs fly away.
How many ladybugs
are left?

+		−
add		subtract

_____ ◯ _____ ◯ _____

ladybug

Problem Solving

Math at Home: Your child solved problems by choosing the operation to write number sentences.
Activity: Put 9 pennies on the table. Then add 3 more. Ask your child to write a number sentence. Then ask your child whether he or she added or subtracted to solve the problem.

Game Zone

2 players

Name_____

Find the Facts!

How to Play:

▶ Take turns. Spin.

▶ Find a fact on the board with that sum or difference.

▶ Put one of your ● on it.

▶ Lose your turn if that sum or difference is covered.

▶ Cover a complete row to win.

You Will Need

10 ●
10 ●

16 −7	8 +8	15 −8	17 −9
8 +5	13 −5	9 +5	8 +7
14 −7	9 +8	10 −5	15 −9

© Macmillan/McGraw-Hill

Technology Link

Fact Families • Computer

- Use 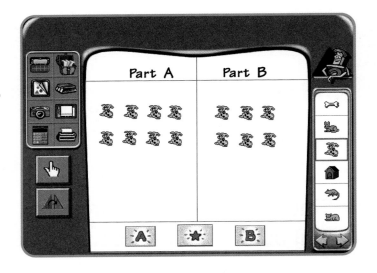 to make a fact family.
- Choose a mat to show two numbers.
- Find the 🐾 stamp.
- Stamp out 8 🐾 and 6 🐾.

Add.	Subtract.
$8 + 6 = \underline{14}$	$14 - 6 = \underline{8}$
$6 + 8 = \underline{\hspace{1cm}}$	$14 - 8 = \underline{\hspace{1cm}}$

8, 6, and 14 make up this fact family.

Complete the fact family.
You can use the computer.

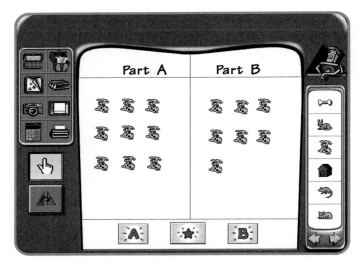

$9 + 7 = \underline{\hspace{1cm}}$

$7 + 9 = \underline{\hspace{1cm}}$

$16 - 9 = \underline{\hspace{1cm}}$

$16 - 7 = \underline{\hspace{1cm}}$

_____, _____, and _____ make up this fact family.

 For more practice use Math Traveler.™

Name_____

Add.
Write the related subtraction facts.

1 7 + 9 = ____

____ ◯ ____ ◯ ____

____ ◯ ____ ◯ ____

2 7 + 6 = ____

____ ◯ ____ ◯ ____

____ ◯ ____ ◯ ____

Complete each fact family.

3 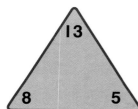 13 / 8 / 5

8 + 5 = ____   13 − 8 = ____

5 + 8 = ____   13 − 5 = ____

4  15 / 9 / 6

9 + 6 = ____   15 − 9 = ____

6 + 9 = ____   15 − 6 = ____

Circle add or subtract.
Write a number sentence to solve.

+	−
add	subtract

5 16 are in the sky.

7 land.

How many are left in the sky? ____ ◯ ____ ◯ ____

© Macmillan/McGraw-Hill

Assessment

Spiral Review and Test Prep
Chapters 1—18

Choose the best answer.

1 Write how many.

10 ⟨⟩ 44 ⟨⟩ 54 ⟨⟩

2 What number is missing in this skip-counting by 2s pattern?

| 8 | 10 | | 14 | 16 |

11 ⟨⟩ 12 ⟨⟩ 15 ⟨⟩

Add.

3 7 + 8 = _____

4 5 + 10 = _____

5 Skip-count by 10s. Write the amount.

_____¢ _____¢ _____¢ _____¢ _____¢ _____¢

6 Show 36¢ two ways. Circle the way that uses fewer coins.

36¢

36¢

THINK
SOLVE
EXPLAIN

LOG ON www.mmhmath.com
For more Review and Test Prep

Telling Time

READ TOGETHER

A Timely Friend

by Charles Ghigna

I have two hands.
I have a face.
I'm found in almost
Every place.

In the kitchen,
In the hall,
I like to hang
Upon the wall.

I like to sit
Beside your bed
Or on the shelf
Near books you've read.

I like to wake
You up each day
And send you off
To school or play.

I like to tick.
I like to tock.
Your timely friend—
I am a CLOCK.

Math at Home

Dear Family,

I will learn to tell and write the time in Chapter 19. Here are my math words and an activity that we can do together.

Love, _____

My Math Words

5 o'clock:

minute hand

hour hand

half past 7:

1 hour is 60 minutes.

1 half hour is 30 minutes.

Home Activity

Have your child find the 1 on a clock face, and read the numbers to 12. Then, point to a number on the clock and ask, "What number comes before?" and "What number comes after?"

© Macmillan/McGraw-Hill

Books to Read

Look for these books at your local library and use them to help your child learn to tell time.

- **Monster Math School Time** by Grace Maccarone, Scholastic, 1997.
- **Isn't It Time?** by Judy Hindley, Candlewick Press, 1996.
- **Telling Time with Big Mama Cat** by Dan Harper, Harcourt Brace and Company, 1998.

LOG ON

www.mmhmath.com
For Real World Math Activities

Name_____

 **Learn** Morning, afternoon, and evening are times of the day.

Math Words

morning

afternoon

evening

before

after

This is what Rob did today.

morning	afternoon	evening

Try It

1 Draw what you do.

morning	afternoon	evening

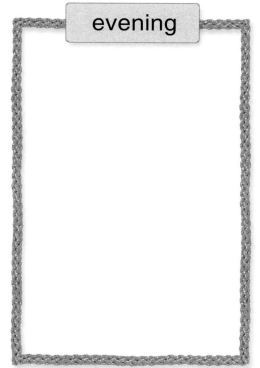

2 ✎ Write **About It!** What do you do every morning?

© Macmillan/McGraw-Hill

Practice Events can happen before and after other events.

Kelly eats lunch.

before

after

Draw what you can do before and after.

3

before school

after school

4

before sleeping

after sleeping

Math at Home: Your child shared what he or she does at certain times of the day.
Activity: Have your child tell you something he or she did at school. Then ask what happened before and after that event.

Name_____

HANDS ON
Activity

Learn The clock has the numbers 1 to 12.

minute hand
hour hand

Math Words

hour hand
minute hand
o'clock

The hour hand is shorter.
It tells the hour.
The minute hand is longer.
It tells the minutes.

The time is 2 o'clock.

Your Turn Use the 🕐 to answer the questions.

1

Where is the hour hand? ___4___

Where is the minute hand? ___12___

2

Where is the hour hand? _____

Where is the minute hand? _____

3 Draw the minute hand to point to the 12.
Draw the hour hand to point to the 11.
Your clock says 11 o'clock.

4 Write **About It!** What is a difference between the
minute hand and hour hand?

© Macmillan/McGraw-Hill

5

Where is the hour hand? __2__

Where is the minute hand? __12__

6

Where is the hour hand? _____

Where is the minute hand? _____

7

Where is the hour hand? _____

Where is the minute hand? _____

8 Draw the minute hand to point to the 12.
Draw the hour hand to point to the 10.
Your clock says 10 o'clock.

9 Draw the minute hand to point to the 12.
Draw the hour hand to point to the 6.
Your clock says 6 o'clock.

 Math at Home: Your child learned how to identify the minute and hour hand on a clock.
Activity: On each hour ask your child to tell you where the minute hand and hour hand are on the clock.

Time to the Hour

 **HANDS ON Activity**

Learn You can use to tell time.

The minute hand is at 12. The hour hand is at 3.

The time is 3 o'clock.

Your Turn Use . Write the time.

1

4 _____ o'clock

2

_____ o'clock

3

_____ o'clock

4

_____ o'clock

5

_____ o'clock

6

_____ o'clock

7 ✏ Write **About It!** The minute hand and the hour hand are at 12. What time is it?

© Macmillan/McGraw-Hill

8

___10___ o'clock

The minute hand is longer. It is at 12. The hour hand is shorter. It tells the hour.

9

_____ o'clock

10

_____ o'clock

11

_____ o'clock

12
_____ o'clock

13

_____ o'clock

14

_____ o'clock

Problem Solving — Visual Thinking

THINK
SOLVE
EXPLAIN

15 Susan goes to the store at 12 o'clock. She comes home one hour later. What time does she come home? Explain.

_____

Math at Home: Your child learned to read and write time to the hour.
Activity: Show your child an analog clock set at 7:00. Then have him or her tell you the time.
Do the same with 4:00 and 12:00.

Time to the Half Hour

HANDS ON
Activity

Learn You can use to tell time to the **half hour**.

The minute hand is at 6.

The hour hand is between 2 and 3.

Math Words

half past
half hour

It is **half past** the hour.
It is half past 2.

Your Turn Use . Write the time.

1

half past **||**

2

half past _____

3

half past _____

4

half past _____

5

half past _____

6

half past _____

7 ✏️ Write **About It!** It is half past 6.
Where is the hour hand on the clock?

© Macmillan/McGraw-Hill

Practice

Use  .
Write the time.

My hour hand is between 12 and 1. My minute hand is at 6. It is half past 12.

⑧

half past _____4_____

⑨

half past _____

⑩

half past _____

⑪

half past _____

⑫

half past _____

⑬

half past _____

Problem Solving — Use Data

Ali's Schedule

Solve.

⑭ What time does Ali wake up?

half past _____

⑮ What time does Ali start school?

_____ o'clock

⑯ What time does Ali go to bed?

half past _____

Wake Up

Start School

Go to Bed

Math at Home: Your child learned to recognize time to the half hour.
Activity: Show your child an analog clock set at 11:30. Have him or her tell you where the hour and minute hands point. Then have him or her tell you the time by using the phrase "half past."

Hour and Half Hour

Learn

Math Words
hour
minutes
half hour

I hour is
60 minutes long.

I half hour is
30 minutes long.

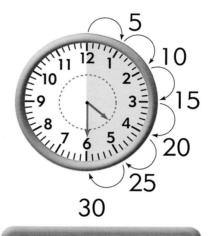

4:00

tells the hour

tells the minutes

It is 4 o'clock.

4:30

It is 30 minutes after 4.
It is half past 4 or 4:30.
4:30 is later than 4:00.

Try It   Write each time. Circle the later time.

1.
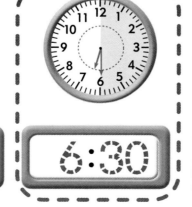

6:00   **6:30**

2.

:   **:**

3. ✏ Write **About It!**   How else can you write 6:30?

© Macmillan/McGraw-Hill

4

1:30 **2:00**

5

: :

6

: :

7

: :

8

: :

9

: :

Problem Solving **Reasoning**

The clocks show when children go to bed.

10 Who goes to bed first?

Alissa Gary Don

_____

11 Who goes to bed last? _____

 Math at Home: Your child learned about hours and half hours.
Activity: Have your child skip-count by fives around an analog clock. Ask your child to tell you the number of minutes in an hour. Then have him or her tell you the number of minutes in a half hour.

Practice Telling Time

Learn

8:00

8:30 is later than 8:00.

8:30

The hour hand is shorter.

Remember, the minute hand is longer.

Try It Draw the hour hand or write the time.

1 3:00

2 2:00

3 11:00

4

:

5

:

6

:

7   Write **About It!** The hour hand is between 1 and 2.
The minute hand is at 6. What time is it?

© Macmillan/McGraw-Hill

Where does the hour hand go? Where does the minute hand go?

8  1:30

9  2:30

10  9:00

11  4:30

12  6:00

13  10:30

Problem Solving **Critical Thinking**

THINK SOLVE EXPLAIN

14 Tina started with this time.
She moved the minute hand
30 minutes.
What time is it now? _____ : _____
Tell how you know.

_____

 Math at Home: Your child practiced telling time to the hour and half hour.
Activity: Show your child an analog watch or clock set to 6:30. Have your child write the time.

Class Visitor

Today Mrs. Cole's class has a different schedule. The children will have Reading and Math, as always. But today a special visitor will come to class!

Morning Schedule

Time	Subject
9:00	Reading
10:00	Math
11:00	Special Visitor 
11:30	Lunch

Problem Solving

Reading Skill **Make Inferences**

1 Who is the special visitor? _____

2 What time will the visitor come to class? _____ : _____

3 How long will the visitor stay?

_____

© Macmillan/McGraw-Hill

Busy Afternoon

The children had a busy afternoon. Mrs. Cole said, "Let us write a letter to thank Captain Smith for visiting." The class liked that idea!

Afternoon Schedule

Time	Subject
12:30	Music
1:00	Science
2:00	Writing
2:30	Library
3:00	Time to go home

Reading Skill

Make Inferences

4 What time will the class write the thank-you letter? _____ o'clock

5 What time does the class go to the library? _____ : _____

6 How long will the class be in the library?

_____

7 What time will the children go home? _____ : _____

 Math at Home: Your child made inferences to answer questions.
Activity: Ask your child how long the class had Music.

Problem Solving Practice

Solve.

1. Andy plays ball at 3:30. Circle the clock that shows 3:30.

THINK SOLVE EXPLAIN ✏ Write a Story!

2. Dee wakes up at 8:00. 2 hours later, she has a drum lesson. What time is her lesson? Write the answer. Then write a story about Dee's day.

___

___

___

---

3. The puppet show starts at 11:00. It ends a half hour later. Draw hands on the clocks to show when the show starts and ends.

Show starts   Show ends

© Macmillan/McGraw-Hill

Writing for Math

 Look at the picture.
Write a story about which
movie the boy can go to.

Show Times
2:00
and
5:00

Writing

Think

What time is it? _____ : _____

What times are the movies? _____ : _____ and _____ : _____

Solve

I can write my story now.

_____

_____

Explain

I can tell you how my answer matches my story.

_____

_____

Name_____

Use the pictures.

1

Making Lunch

When did this happen?
Circle your answer.

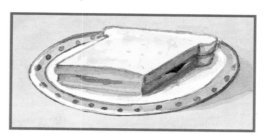

before **or** after

Write each time.

2

_____ o'clock

3

half past _____

Write each time. Circle the later time.

4

5 How else can you write
half past 2?   2:00   2:30

© Macmillan/McGraw-Hill

Assessment

Spiral Review and Test Prep

Chapters 1–19

Choose the best answer.

1 Tyrell has these coins. What can he buy?

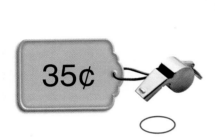
35¢
○

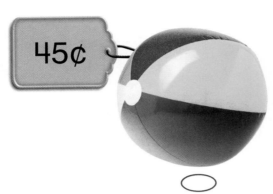
45¢

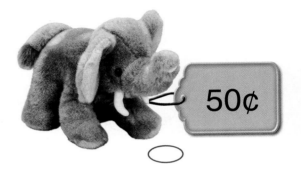
50¢
○

○

---

2 Which number comes between?

| 22, _____, 24 |

21   23   25
○   ○   ○

Write how many.

3

tens	ones

_____

4

tens	ones

_____

THINK SOLVE EXPLAIN

5 Use 3, 5, and 8 to show a fact family.

_____

Time and Calendar

New Bicycle

Sunday I got a brand-new bike;

Monday I learned how to ride;

Tuesday I went by my grandmother's house,

And to the countryside.

Wednesday I pedaled up a hill;

Thursday I reached the top;

I'll be home Friday or Saturday—

Or as soon as I learn how to stop.

by Yolanda Nave

Math at Home

Dear Family,

I will learn about time and calendars in Chapter 20. Here are my math words and an activity that we can do together.

Love, _____

My Math Words

month

yesterday

today

tomorrow

Yesterday was March 6. Today is March 7. Tomorrow will be March 8.

Home Activity

Discuss with your child daily activities that seem to take a long time or a short time.

You can make a chart.

Short Time	Long Time
• getting out of bed	• sleeping at night
• brushing teeth	• being at school
• putting on shoes	

© Macmillan/McGraw-Hill

Books to Read

In addition to these library books, look for the Time for Kids math story that your child will bring home at the end of this unit.

- **What Time Is It?** by Sheila Keenan, Scholastic, 1999.
- **Melody Mooner Takes Lessons** by Frank B. Edwards, Bungalo Books, 1996.
- **Time for Kids**

Party!
I am having a party.
I am asking 14 friends to come.
Will you help me plan?

LOG ON
www.mmhmath.com
For Real World Math Activities

Learn A clock and a calendar measure time in different ways.

Math Words

calendar

month

days of the week

year

A clock measures minutes and hours.

A calendar measures days of the week , weeks , months , and years . The months and days of the week are at the top of the calendar.

8:30

June

S	M	T	W	T	F	S
				1	2	3	4
5	6	7	8	9	10	11
12	13	14	15	16	17	18
19	20	21	22	23	24	25
26	27	28	29	30		

Try It Circle your answers.

	Things you do	About how long would it take?	How would you measure?
1	play a ball game	1 minute (hour) day	(clock)	**May**
2	tie a shoelace	1 minute hour day	(clock)	**June**
3	grow a new tooth	3 minutes days months	(clock)	**September**

4 ✏ **Write About It!** Write something you can do in about 1 hour.

© Macmillan/McGraw-Hill

Practice Circle your answers.

A clock measures shorter times.

A calendar measures longer times.

Things you do	About how long would it take?	How would you measure?
⑤ make lunch	10 **(minutes)** weeks months	clock (circled) / April calendar
⑥ watch a movie	2 minutes hours days	clock / August calendar
⑦ grow a tall tree	15 days weeks years	clock / November calendar

Problem Solving — Estimation

About how long would it take?
Would it happen in the morning, afternoon, or evening?

⑧

| 10 minutes | 10 days |

| morning | afternoon | evening |

⑨

| 8 minutes | 8 hours |

| morning | afternoon | evening |

 Math at Home: Your child estimated lengths of time and worked with instruments that measure time.
Activity: Talk about keeping track of time with your child. Ask which is longer: one month or one hour; one minute or one year; one week or one day. Talk about activities that take minutes, days, months, and years to complete.

Name_____

Write the number that comes between.

41 __42__ 43

25 ____ 27

80 ____ 82

34 ____ 36

16 ____ 18

78 ____ 80

11 ____ 13

62 ____ 64

93 ____ 95

50 ____ 52

74 ____ 76

35 ____ 37

22 ____ 24

© Macmillan/McGraw-Hill

Count how much money is in each bank.

__12__ ¢

_____ ¢

_____ ¢

_____ ¢

_____ ¢

_____ ¢

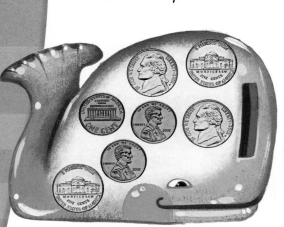

_____ ¢

_____ ¢

_____ ¢

LOG ON www.mmhmath.com
For more practice

Math at Home: Your child has practiced counting money.
Activity: Show your child different amounts of coins and ask him or her to tell you how much you have.

Name_____

Learn A calendar tells months, days, and dates. This calendar shows 1 month. It can also show you yesterday , today , and tomorrow .

Math Words

yesterday
today
tomorrow

May

Sunday	Monday	Tuesday	Wednesday	Thursday	Friday	Saturday
1	2	3	4	5	6	7
8	9	10	11	12	13	14
15	16	17	18	19	20	21
22	23	24	25	26	27	28
29	30	31				

There are 7 days in 1 week.

Try It Use the calendar to answer the questions.

1 What month does the calendar show?

May

2 On what day of the week does the month begin?

_____

3 Today's date is May 6. What day of the week is tomorrow?

_____

4 Today is May 26. What day of the week was yesterday?

_____

5 ✏️ Write **About It!** How can you tell when the month ends?

© Macmillan/McGraw-Hill

This calendar shows one year.
There are 12 months in one year.
January is the first month of the year.

6 Which is the third month? _____March_____

7 How many Tuesdays are there in the month of October? _____

8 Which is the tenth month? _____

9 How many days are there in September? _____

10 Which month comes just after June? _____

Math at Home: Your child used a calendar.
Activity: Use the calendar on the top of the page. Ask your child to show you the third, fifth, and ninth months
of the year. Then find special days like birthdays and mark them on the calendar with your child.

Name _____

Problem Solving Strategy

Find a Pattern • Algebra

You can find a pattern to help you solve problems.

Casey made a ladybug pattern on the calendar. The ladybugs are on the odd numbers. Which dates would continue this pattern?

April

Sunday	Monday	Tuesday	Wednesday	Thursday	Friday	Saturday
					1	2
3	4	5	6	7	8	9
10	11	12	13	14	15	16
17	18	19	20	21	22	23
24	25	26	27	28	29	30

Problem Solving

Read

What do I already know? _____

What do I need to find? _____

Plan

I need to find the next odd numbers to continue the pattern.

Solve

I can carry out my plan.

_____ and _____ are the next numbers in the pattern.

Look Back

Does my answer make sense?   Yes.   No.

How do I know? _____

© Macmillan/McGraw-Hill

Find the patterns to answer the questions.

June

Sunday	Monday	Tuesday	Wednesday	Thursday	Friday	Saturday
			1	2	3	4
5	6	7	8	9	10	11
12	13	14	15	16	17	18
19	20	21	22	23	24	25
26	27	28	29	30		

Problem Solving

1. Find the 🦗 on the calendar. What is the pattern?

_____

How do you know?

_____

2. Look at the 🐝. Which dates would the next 🐝 be on? _____

How do you know?

_____

3. How are June 8 and June 15 alike?

_____

How do you know?

_____

Math at Home: Your child looked for patterns on a calendar.
Activity: For one month have your child draw a sun or a cloud to represent each day's weather on a calendar. Ask him or her to look for patterns in the weather.

Game Zone

Calendar Chase

How to Play:

▶ Take turns. Toss the .

▶ Move your ⚪ that number of spaces. Name that date.

▶ Pick a card. Go to that day.

▶ The first person to reach November 30 wins.

👥 **2 players**

Yesterday

Today

Tomorrow

You Will Need

1 🎲

2 ⚪

3 word cards

November

Sunday	Monday	Tuesday	Wednesday	Thursday	Friday	Saturday
		start 1	2	3	4	5
6	7	8	9	10	11	12
13	14	15	16	17	18	19
20	21	22	23	24	25	26
27	28	29	finish 30			

© Macmillan/McGraw-Hill

Technology Link

Skip-Count Time • Calculator

You Will Use

You can use a  to skip-count to 60 minutes.
Skip-count by 15.

Press. (On/Off) [0]　　　　　　*0*

[+] [1] [5] [=]　　　　　　*15*

[+] [1] [5] [=]　　　　　　*30*

[+] [1] [5] [=]　　　　　　*45*

[+] [1] [5] [=]　　　　　　*60*　　*60* minutes

---

1 Skip-count by 6 to 30 minutes.

Press. (Clear) [0]　　　　　*0*

[+] [6] [=]　　　　　*6*

[+] [6] [=]

[+] [6] [=]

[+] [6] [=]

[+] [6] [=]

_____ minutes

2 Skip-count by 12 to 60 minutes.

Press. (Clear) [0]　　　　　*0*

[+] [1] [2] [=]　　　　*12*

[+] [1] [2] [=]

[+] [1] [2] [=]

[+] [1] [2] [=]

[+] [1] [2] [=]

_____ minutes

 For more practice use Math Traveler.™

360 three hundred sixty

Name

June

Sunday	Monday	Tuesday	Wednesday	Thursday	Friday	Saturday
			1	2	3	4
5	6	7	8	9	10	11
12	13	14	15	16	17	18
19	20	21	22	23	24	25
26	27	28	29	30		

Use this calendar to answer the questions.

1. How many Fridays are there in June? _____

2. What day of the week is June 26? _____

3. Today is June 23. What day of the week was yesterday?

   _____

4. Today is June 11. What day
   of the week is tomorrow? _____

5. Write the number that comes next in the pattern.

   7 6 6 7 6 6 6 7 6 6 6 7 6 _____

© Macmillan/McGraw-Hill

Assessment

Choose the best answer.

1 7 + 3 + 5 = _____

13 ⬭   14 ⬭   15 ⬭

---

2 I am an odd number.
I am between 56 and 60.
I am not 59.
What number am I?

55 ⬭   57 ⬭   58 ⬭

---

3 How many more children like ladybugs than crickets?

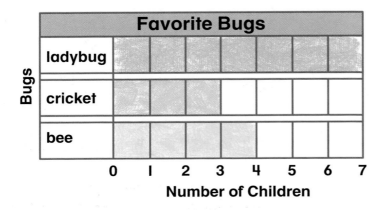

Favorite Bugs

3 ⬭   4 ⬭   5 ⬭

---

4 The clocks show the times the children wake up. Who wakes up first?

LeeAnn   Chun

_____

THINK
SOLVE
EXPLAIN

5 Peter has 45¢ in his bank. What coins might he have?

_____

Name _____

1 of my 9 friends ate 2 cupcakes.

The rest of my friends ate 1 cupcake each.

Color in how many cupcakes they ate in all.

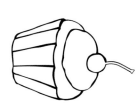
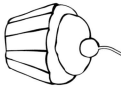
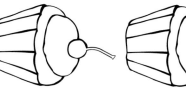
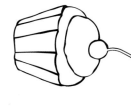

Fold down

TIME FOR KIDS

Party!

I am having a party.
I am asking 14 friends to come.
Will you help me plan?

© Macmillan/McGraw-Hill

READ TOGETHER

Party Time!

Day:
Saturday

Time:
1:00 in the afternoon

Please let me know by:
Thursday

I am sending out 14 invitations.
I have only 6 stamps.
I need 8 more stamps.

$$\begin{array}{r} 14 \\ -\ 6 \\ \hline 8 \end{array}$$

---

9 of my friends said they can come.
5 of my friends cannot come.

$$\begin{array}{r} 14 \\ -\ 9 \\ \hline 5 \end{array}$$

---

9 friends are coming.
The doorbell rings.
8 of my friends are here.
I am still waiting for 1 friend.

$$\begin{array}{r} 9 \\ -\ 8 \\ \hline 1 \end{array}$$

Linking Math and Science

Calendar and the Moon

© Macmillan/McGraw-Hill

Science Word

moon

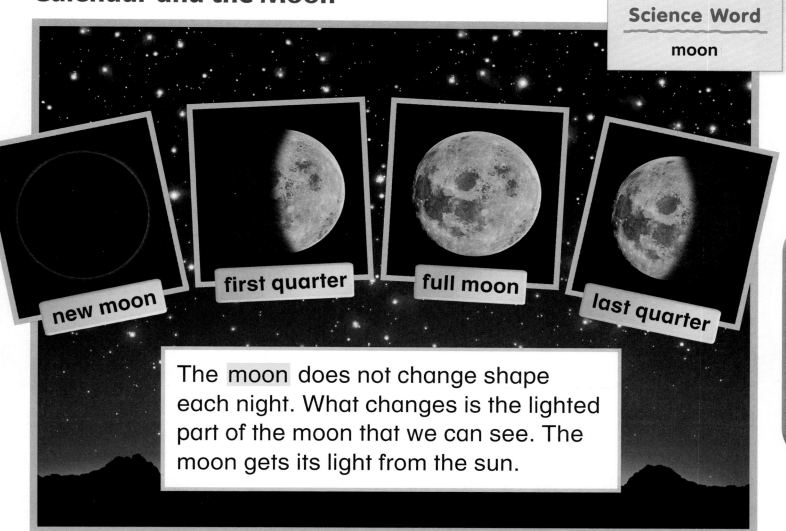

new moon

first quarter

full moon

last quarter

The moon does not change shape each night. What changes is the lighted part of the moon that we can see. The moon gets its light from the sun.

Problem Solving

Look at the picture.
Circle the word to complete each sentence.

1 The moon gets its light from the _____.

    earth   sun

2 Does the moon change its shape?

    Yes.   No.

May

Sunday	Monday	Tuesday	Wednesday	Thursday	Friday	Saturday
1 new moon	**2**	**3**	**4**	**5**	**6**	**7** quarter moon
8	**9**	**10**	**11**	**12**	**13**	**14**
15 full moon	**16**	**17**	**18**	**19**	**20**	**21**
22 quarter moon	**23**	**24**	**25**	**26**	**27**	**28**
29 new moon	**30**	**31**				

Problem Solving

Use the calendar to answer the questions.

1 Which date does the  fall on?

_____

2 Which day of the week does the 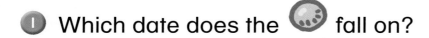 fall on?

_____

3 Which day of the week comes before 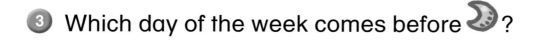?

_____

4 Which dates does the 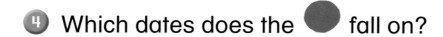 fall on?

_____

 Math at Home: Your child used a calendar to investigate phases of the moon.
Activity: Help your child figure out how many full moon phases there are in one year.

Name _____

Math Words

Draw lines to match.

1 6:30

2 60 minutes

3 6 + 6

doubles

I hour

half past 6

Skills and Applications

Addition and Subtraction Facts to 20 (pages 293–295, 313–322)

Examples

> Doubles and doubles plus I facts can help you add.
>
> $6 + 6 = 12$
> $6 + 7 = 13$

4 $7 + 7 = $ ____

$7 + 8 = $ ____

5 $5 + 5 = $ ____

$5 + 6 = $ ____

> Use addition doubles to subtract.
>
> $8 + 8 = 16$
> $16 - 8 = 8$

6 $2 + 2 = $ ____

$4 - 2 = $ ____

7 $10 + 10 = $ ____

$20 - 10 = $ ____

> Use related subtraction facts.
>
> $9 - 7 = 2$
> $9 - 2 = 7$

8 $11 - 4 = $ ____

$11 - 7 = $ ____

9 $10 - 6 = $ ____

$10 - 4 = $ ____

© Macmillan/McGraw-Hill

Skills and Applications

Telling Time (pages 331–342)

Examples

You can tell time to the half hour. The hour hand is between the 2 and the 3. The minute hand is on the 6.

The time is half past 2, or 2:30

10

half past _____

11

half past _____

(pages 323–324, 357–358)

Problem Solving — Strategy

Write a number sentence to solve. Draw a picture to check.

Mom baked 8 muffins. She is baking 6 more. How many muffins will there be in all?

$$8 + 6 = 14 \text{ muffins}$$

12 Molly ate 9 grapes. Then Molly ate 7 more grapes. How many grapes did Molly eat?

____ ◯ ____ ◯ ____

Math at Home: Your child learned addition and subtraction strategies and telling time. Have your child use these pages to review.
Activity: Work with your child at home to tell time on the hour and the half hour.

Name_____

Using Time

Draw the clock hands to show the time.
Draw a picture of something you might
do at each time.

8:30

3:30

3

12:00

 You can put this page in your portfolio.

© Macmillan/McGraw-Hill

Assessment

Unit 5
Enrichment

Elapsed Time

Jack went to the butterfly show at 4:00.

1 Draw clock hands to show when he goes to the show.

The show was over 2 hours later. Jack leaves when it was over.

2 Draw clock hands to show when Jack leaves.

Jack takes the bus home. He gets home in 30 minutes.

3 Draw clock hands to show what time Jack gets home.

Jack eats dinner 1 hour after he gets home.

4 Draw clock hands to show what time Jack eats dinner.

SANDCASTLES
EVERYWHERE

READ TOGETHER

Story by Jane Ito
Illustrated by Jennifer Rarey

Sandcastles, sandcastles everywhere!
We can make one over there.
Get ready, get set!
We first get wet!

Circle the animal that can go the highest.

© Macmillan/McGraw-Hill

Which sunblock will he use?
It is so hard to choose.

Circle the tallest bottle.

© Macmillan/McGraw-Hill

Crab will lend a hand.
He will dig in the sand.

Circle the shortest shovel.

Sandcastles everywhere.
Over here and over there.

Circle the tallest castle.
Put an X on the shortest castle.

© Macmillan/McGraw-Hill

Math at Home

Dear Family,

I will learn to estimate and measure length in Chapter 21. Here are my math words and an activity that we can do together.

Love, _____

My Math Words

inch :

A customary unit of measurement about two finger spaces long.

 ← **I inch**

foot :

I foot = 12 inches

centimeter :

A metric unit of measurement about one finger space long.

← **I centimeter**

Home Activity

Cut 4 straws or paper strips into different lengths. Have your child arrange the straws from shortest to longest.

Challenge your child to find objects around your home that match the lengths of the straws.

 www.mmhmath.com
For Real World Math Activities

Books to Read

Look for these books at your local library and use them to help your child learn about length.

- **Much Bigger Than Martin** by Steven Kellogg, Dial Books, 1976.
- **The Long and Short of It** by Cheryl Nathan and Lisa McCourt, BridgeWater Books, 1998.
- **Inch by Inch** by Leo Lionni, Mulberry Books, 1960.

© Macmillan/McGraw-Hill

Explore Length

HANDS ON
Activity

Learn You can **measure** to find how long something is. You can use cubes to measure.

Math Words

measure
longest
shortest

The red ribbon is about 6 cubes long. It is the **longest** ribbon. The **shortest** ribbon is green.

Your Turn Estimate about how many long. Then use to measure.

1.

Estimate: about ___6___   Measure: about ___6___

2.

Estimate: about _____   Measure: about _____

3.

Estimate: about _____   Measure: about _____

4. ✎ Write **About It!** Which ribbon is longest? Which ribbon is shortest? Explain.

© Macmillan/McGraw-Hill

Practice
Estimate how many 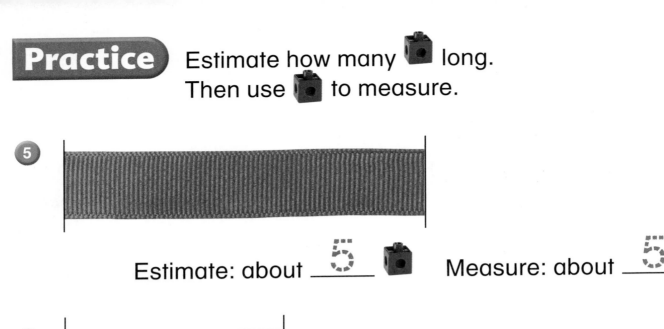 long.
Then use 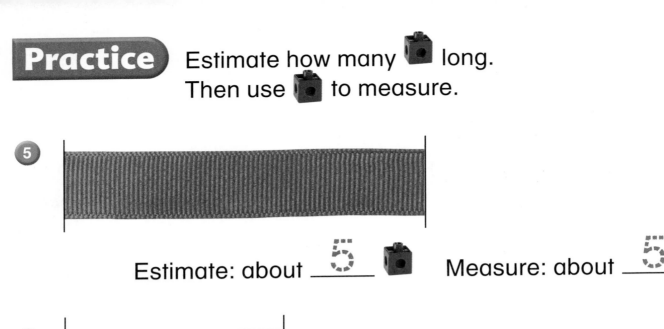 to measure.

5

Estimate: about __5__ 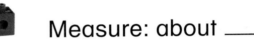   Measure: about __5__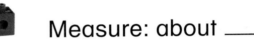

6

Estimate: about _____   Measure: about _____

7

Estimate: about _____ 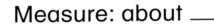   Measure: about _____

Problem Solving | Reasoning

THINK
SOLVE
EXPLAIN

8 Tim used  to measure a ribbon.
Kim used  to measure the same ribbon.

Which is true? Circle it.

A Tim used more .

B Kim used more .

C They both used the same number of  and .

Math at Home: Your child used cubes to measure length.
Activity: Have your child use paper clips or toothpicks to measure and compare the lengths of objects.

Name_____

Inch

Learn You can use an inch ruler to measure.

Math Word
___
inch

This fish is about 4 inches long.

2 fingers are about 1 inch wide.

Your Turn Estimate how long. Then use an inch ruler to measure. Write how many inches.

1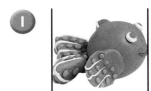

Estimate: about _____ inch   Measure: about _____ inch

2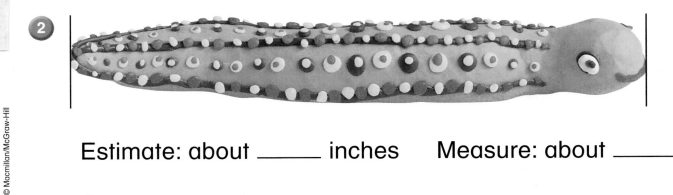

Estimate: about _____ inches   Measure: about _____ inches

3 ✎ Write **About It!** Would your measures be the same if you used a large paper clip to measure? Explain.

© Macmillan/McGraw-Hill

Rincker Memorial Library-Concordia

Practice Find these objects in your classroom. Estimate how long. Then use an inch ruler to measure.

4

Estimate: about _____ inches

Measure: about _____ inches

5

Estimate: about _____ inches

Measure: about _____ inches

6  Glue Stick

Estimate: about _____ inches

Measure: about _____ inches

7

Estimate: about _____ inches

Measure: about _____ inches

8

Estimate: about _____ inches

Measure: about _____ inches

9

Estimate: about _____ inches

Measure: about _____ inches

Problem Solving — Estimation

10 The blue pencil is 3 inches long.
About how long is the green pencil?

about _____ inches long

Math at Home: Your child used an inch ruler to measure length to the nearest inch.
Activity: Have your child use an inch ruler to measure small objects around your home, such as fruits and vegetables. Have him or her arrange the objects from shortest to longest.

Inch and Foot

Learn There are 12 inches in 1 foot.

You can use inches to measure short objects. You can use feet to measure long objects.

Math Word
_____
foot

Try It Which is better for measuring the real object?
Circle inches or feet.

1

inches   (feet)

2

inches   feet

3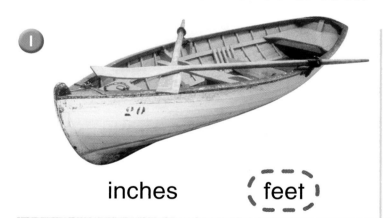

inches   feet

4

inches   feet

5 Write **About It!** If a shell is 1 inch long and a fish is
1 foot long, which is longer? Explain.

© Macmillan/McGraw-Hill

Which is better for measuring the real object? Circle inches or feet.

 Use inches for measuring short objects. Use feet for measuring long objects.

6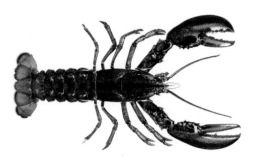

(inches)　　feet

7

inches　　feet

8

inches　　feet

9

inches　　feet

10

inches　　feet

11

inches　　feet

Problem Solving　**Number Sense**

 Show Your Work

THINK
SOLVE
EXPLAIN

12 The green fish swims 7 feet.
The spotted fish swims 3 feet.
How many more feet does the
green fish swim?

_____ feet

 Math at Home: Your child learned when to use inches and feet for measuring length.
Activity: Point out objects during a walk around your neighborhood, and have your child tell if it is better to use inches or feet to measure them.

Name_____

Understanding Measurement

Learn You can use a ruler to draw and measure.

My string is 5 inches long. I start at 0 and end at 5.

HANDS ON Activity

Your Turn Use an inch ruler and a crayon to draw and measure.

1. Draw a **blue** string that is 2 inches long.

2. Draw a **red** string that is 3 inches long.

3. Draw a **green** string that is 6 inches long.

4. Connect the dots to draw a **purple** line.

   •   •

   Estimate how long.   about _____ inches

   Measure how long.   about _____ inches

5. Write **About It!** How long is a string that is 1 inch longer than 6 inches?

© Macmillan/McGraw-Hill

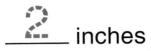
Use an inch ruler to draw and measure.

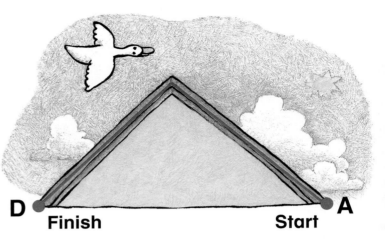

6 Measure how long from A to B.

2 _____ inches

7 Measure how long from B to C.

_____ inches

8 Measure how long from C to D.

_____ inches

9 Estimate from D to A.

_____ inches

Measure: _____ inches

Problem Solving ⟨ **Estimation** ⟩

 THINK SOLVE EXPLAIN

10 The top of this box is 2 inches long. About how long are the other sides? Estimate.

blue: about _____ inches

green : about _____ inches

purple : about _____ inches

Then use a ruler to check.

 Math at Home: Your child learned how to draw and measure lines using an inch ruler.
Activity: Have your child draw straight lines to make a shape, such as a house or a robot.
Have him or her estimate the length of each line and then measure to check.

Name_____   **Centimeter**

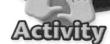

Learn You can use a  centimeter ruler to measure.

This turtle is about 4 centimeters long.

centimeters

I finger is about I centimeter wide.

Your Turn Estimate how long. Then use a centimeter ruler to measure. Record.

Estimate: about _____ centimeters   Measure: __5__ centimeters

Estimate: about _____ centimeters   Measure: _____ centimeters

Estimate: about _____ centimeters   Measure: _____ centimeters

 ✎ Write **About It!** How can you draw a string that is 9 centimeters long?

© Macmillan/McGraw-Hill

Find these things in your classroom. Estimate how long. Then use a centimeter ruler to measure.

You can use your finger to estimate a centimeter.

Find	Estimate	Measure
5	about _____ centimeters	about _____ centimeters
6	about _____ centimeters	about _____ centimeters
7	about _____ centimeters	about _____ centimeters
8	about _____ centimeters	about _____ centimeters

Make it Right

THINK
SOLVE
EXPLAIN

9 This is how Liz measured her eraser.

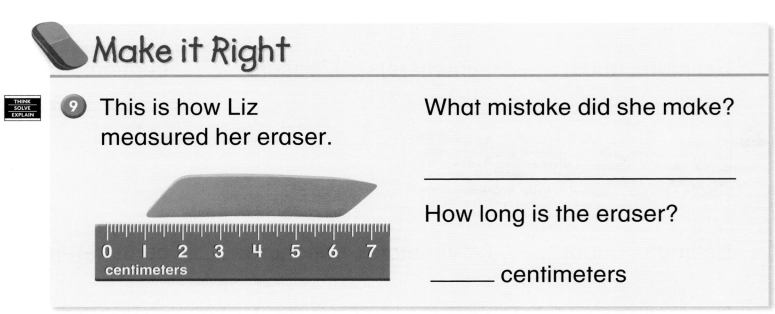

What mistake did she make?

_____

How long is the eraser?

_____ centimeters

Math at Home: Your child used a centimeter ruler to measure length to the nearest centimeter.
Activity: Measure the centimeter lengths of three objects. Record the lengths on separate paper strips. Ask your child to match the objects with the measurements. Then have your child measure with a centimeter ruler to check.

Sand Castles

Jon and Tina build sand castles.
Jon says his castle is taller.
Tina says hers is taller.
Dad says, "There is a
way we can find out!"

Problem Solving

Reading Skill **Predict Outcomes**

1 Which castle do you think is taller?

_____

2 What will Dad do to find out?

_____

3 Jon's castle is about 4 tall. Tina's is about 3 tall.

Which castle is taller? _____

4 How much taller? about _____ taller

© Macmillan/McGraw-Hill

Footprints

Jon and Tina make footprints.
"I can measure the footprints with
my tape measure," says Mom.
Jon's footprint is 5 inches long.
Tina's footprint is 7 inches long.

Problem Solving

Reading Skill **Predict Outcomes**

5 How much longer is Tina's footprint? _____ inches

6 Mom measures her footprint. It is 8 inches long.

Mom's footprint is _____ inches longer than Jon's.

7 Do you think Dad's footprint would be longer or shorter

than Jon's? _____

8 What do you think Tina and Jon might measure next?

_____

_____

Math at Home: Your child predicted outcomes to answer questions.
Activity: Show your child two items from around the house and have him or her predict which is longer or taller.
Then ask your child to measure the items with a ruler or a piece of string to determine which truly is longer or taller.

Problem Solving Practice

Solve.

THINK SOLVE EXPLAIN ✏️ **Write a Story!**

1 Ken's 🐚 is 3 inches long.
Ann's 🐚 is 2 inches longer.
Who has the longer 🐚? Explain.

_____

_____

2 Dave's sand castle is 7 inches tall.
Peter's sand castle is 3 inches shorter than Dave's.
Anita's sand castle is 4 inches taller than Dave's.
How tall are Peter's and Anita's sand castles?

Peter _____ inches tall   Anita _____ inches tall

Which sand castle is almost 1 foot tall? _____

3 Jill brought a pail, shovel, and rake
to the beach. The pail is 9 inches tall.
The shovel is 4 inches taller than the pail.
The rake is 4 inches shorter than the pail.

How tall? shovel _____ rake _____

Which object is taller than 1 foot? _____

© Macmillan/McGraw-Hill

Problem Solving

Writing for Math

 Write a math story about the picture.
Use a ruler to compare the fish.
Here are some words to help you.

longer

shorter

inch

Writing

Think

Should I use feet or inches to measure the fish?

_____

Solve

I can write my math story now.

_____

_____

Explain

I can tell you how my story matches the picture.

_____

_____

Name_____

1 Use  to measure.

about _____

2 Use an inch ruler to measure.

about _____ inches

3 Use a centimeter ruler to measure.

about _____ centimeters

4 Estimate how long. Use an inch ruler to measure.

Estimate: about _____ inches   Measure: about _____ inches

5

Estimate: about _____ inches   Measure: about _____ inches

© Macmillan/McGraw-Hill

Assessment

Spiral Review and Test Prep
Chapters 1–21

Choose the best answer.

1 Which day of the week comes just after Monday?

Tuesday ⬭   Thursday ⬭   Sunday ⬭

---

2 If today is March 12, what date was yesterday?

February 12 ⬭   March 11 ⬭   March 13 ⬭

---

3 $16 - 8 = \blacksquare$

9 ⬭   8 ⬭   6 ⬭

---

4 Draw the clock hands. Write the time.
Roger starts to read at 7:00. He stops 30 minutes later.

 Roger starts, at ___ : ___ .

 Roger stops, at ___ : ___ .

THINK SOLVE EXPLAIN

5 $6 + 7 = 13$ and $13 - 7 = 6$ belong to the same fact family.
Tell how you know.

_____

_____

READ TOGETHER

Homework

A bowl of rice is heavy.
A bowl of socks weighs less.

I like to weigh things all day long.
I like to make a mess.

387

Math at Home

Dear Family,

I will learn ways to estimate and measure weight, capacity, and temperature in Chapter 22. Here are math words and an activity that we can do.

Love, _____

My Math Words

heavier , lighter :

An  is heavier than a paperclip.

A paperclip is lighter than an .

temperature :

You use temperature to measure how hot or cold.

degrees :

Use degrees to measure temperature.

The temperature is 50 degrees.

Home Activity

Show your child two different-sized bowls. Ask your child which bowl would hold less water. Help your child fill that container with water.

Then help your child test his or her answer by pouring the water into the other bowl.

© Macmillan/McGraw-Hill

Books to Read

Look for these books at your local library and use them to help your child learn about capacity, weight, and temperature.

- **Temperature and You** by Betsy Maestro and Giulio Maestro, Lodestar Books, 1990.
- **What's Up with That Cup?** by Sheila Keenan, Scholastic, 2000.
- **The 100-Pound Problem** by Jennifer Dussling, The Kane Press, 2000.

LOG ON
www.mmhmath.com
For Real World Math Activities

Name_____

Explore Weight

Learn You can use a  to see which objects are <mark>heavier</mark> and which are <mark>lighter</mark>.

The object that goes down is heavier. The feather is lighter than the block.

Math Words
heavier
lighter

Your Turn Use a  to compare each pair of objects. Circle the heavier one.

1.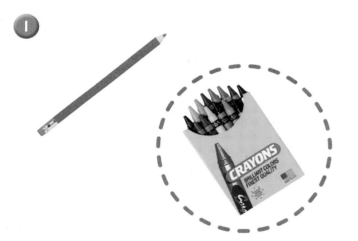

2.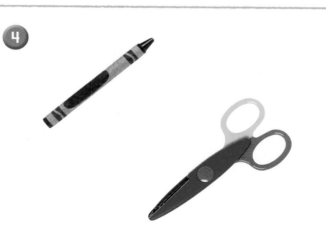

3.

4.

© Macmillan/McGraw-Hill

5. ✎ Write **About It!** How can you tell which object is heavier?

Practice  Compare each object to a 🔲.
Circle your estimate. Then use a ⚖
to measure. Circle your answer.

Object	Estimate	Measure
⑥ CRAYONS BRILLIANT COLORS FINEST QUALITY	(heavier) lighter	(heavier) lighter
⑦	heavier lighter	heavier lighter
⑧	heavier lighter	heavier lighter

Problem Solving ◖ Estimation

Do you think the container holds more
or less than the bottle? Circle.

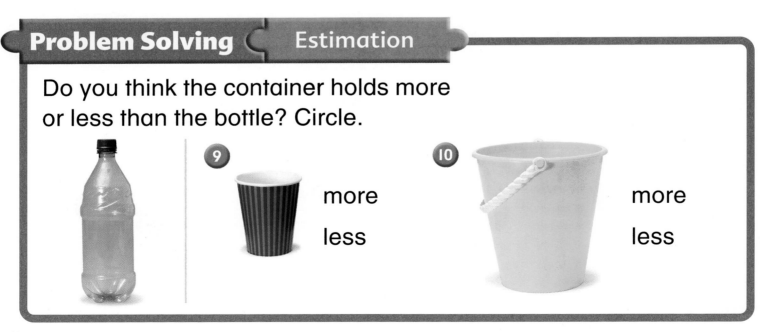

⑨ more

less

⑩ more

less

 Math at Home: Your child estimated which objects are heavier or lighter than others.
Activity: Show a book and a piece of fruit to your child. Ask him or her to tell you which object
is heavier. Do the same with other objects around your home.

Name_____   **Cup, Pint, Quart**

Learn You can use cups, pints, and quarts to measure how much a container will hold.

2 cups = one pint
2 pints = one quart

Math Words

cup
pint
quart

I quart   I pint   I cup

Your Turn Circle the one that holds the same amount. You can measure to check.

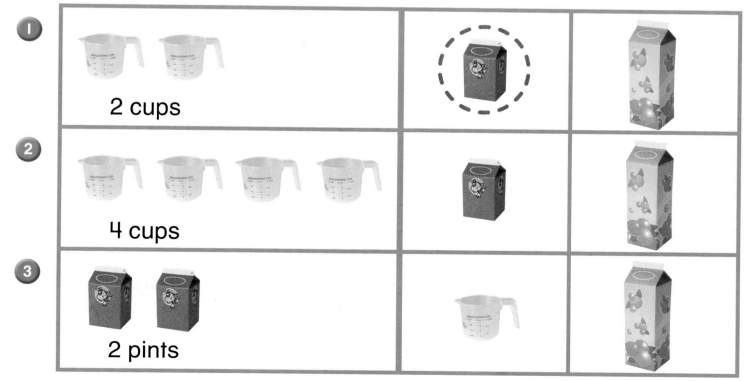

1 2 cups

2 4 cups

3 2 pints

4 ✎ **Write About It!** How many cups of juice would fill a quart container? Explain.

© Macmillan/McGraw-Hill

Practice Compare each container to the .
Circle the best estimate.
You can measure to check.

> A pint is more than a cup.
> A quart is more than a pint.

Container	Compare	Estimate
5	I cup	less than I cup / more than I cup
6	I pint	less than I pint / more than I pint
7	I pint	less than I pint / more than I pint
8	I cup	less than I cup / more than I cup
9	I quart	less than I quart / more than I quart

Math at Home: Your child measured using cups, pints, and quarts.
Activity: Have your child find a container in your home that holds about a pint and explain how he or she knew.

Name_____ **Pound**

Learn You can measure weight in pounds.

This apple is less than I pound.

Math Word

pound

about
I pound

more than
I pound

Your Turn Estimate. Circle your answer.
Then measure with a and a  .

Object	Estimate	Measure
① paper clip	*less than I pound* more than I pound	*less than I pound* more than I pound
② stapler	less than I pound more than I pound	less than I pound more than I pound

③ ✏ Write **About It!** A kitten weighs about I pound. Does a cat weigh more or less than I pound? How do you know?

© Macmillan/McGraw-Hill

Practice   Estimate. Circle your answer.
Then measure with a 🔋 and a ⚖️.

Object	Estimate	Measure
④ Glue Stick	(less than I pound) more than I pound	(less than I pound) more than I pound
⑤ (soccer ball)	less than I pound more than I pound	less than I pound more than I pound
⑥ (scissors)	less than I pound more than I pound	less than I pound more than I pound
⑦ DINOSAURS (book)	less than I pound more than I pound	less than I pound more than I pound

Problem Solving   Mental Math

⑧   Mica has 3 pounds of 🍎.
He gets 2 more pounds.
How many pounds of 🍎 does
he have in all?

_____ pounds

Math at Home: Your child measured objects using pounds.
Activity: Have your child find objects in your home that might weigh less than one pound.
Then have him or her find objects that might weigh more than one pound.

Name_____ **Liter**

Learn You can also use liters to measure how much a container holds.

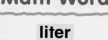
Math Word

liter

less than
I liter

I liter

more than
I liter

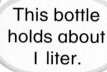

This bottle holds about I liter.

Your Turn Compare each container to I liter.
Circle the best estimate.
You can measure to check.

Container	Estimate	Container	Estimate
①	less than I liter (more than I liter)	②	less than I liter more than I liter
③	less than I liter more than I liter	④	less than I liter more than I liter

⑤ Write **About It!** How can you find out if a cup holds more or less than I liter?

© Macmillan/McGraw-Hill

Estimate. If the container holds less than 1 liter, write the word **less**. If it holds more than 1 liter, write the word **more**.

1 liter

6

7

8

Problem Solving **Use Data**

Use the graph.

9 Which container holds 4 liters?

10 How many more liters does the hold than the ?

_____ more liters

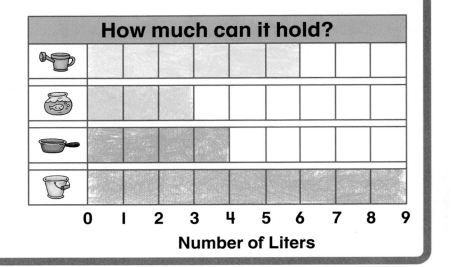

How much can it hold?

0 1 2 3 4 5 6 7 8 9
Number of Liters

Math at Home: Your child measured using liters.
Activity: Have your child find three containers in your home that hold more than one liter.

Name_____

Learn You can use grams to measure light objects. You can use kilograms to measure heavy objects.

Math Words

gram

kilogram

The pumpkin is heavy. Use kilograms to measure it.

The peach is light. Use grams to measure it.

Try It Circle the better unit of measure.

①

(grams)

kilograms

②

grams

kilograms

③

grams

kilograms

④

grams

kilograms

⑤

grams

kilograms

⑥

grams

kilograms

⑦ **Write About It!** Would you use grams or kilograms to measure a mouse? Why?

© Macmillan/McGraw-Hill

Use grams to measure light objects. Use kilograms for heavy objects.

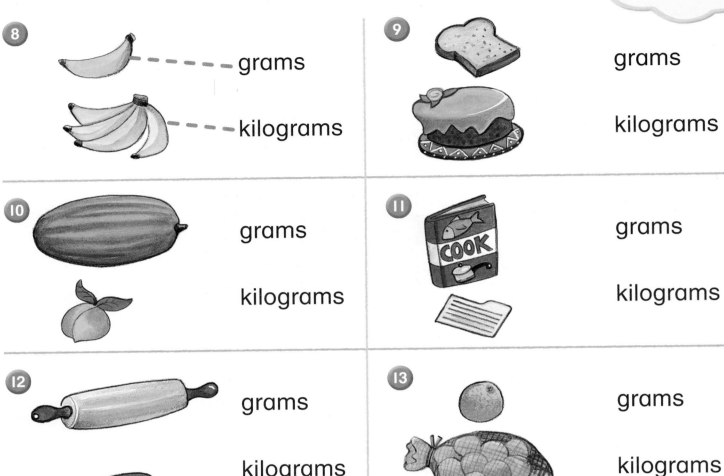

8
grams

kilograms

9
grams

kilograms

10
grams

kilograms

11
grams

kilograms

12
grams

kilograms

13
grams

kilograms

Spiral Review and Test Prep

Choose the best answer.

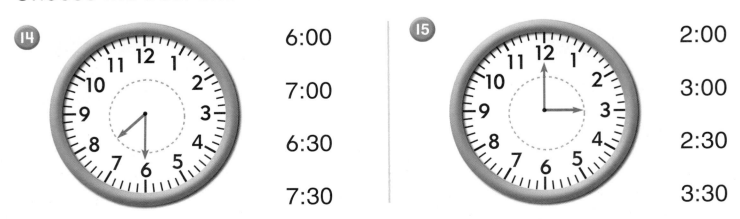

14
6:00

7:00

6:30

7:30

15
2:00

3:00

2:30

3:30

Math at Home: Your child has learned when to use grams or kilograms to measure objects.
Activity: Show your child a variety of light and heavy objects. Ask him or her which should be measured by grams and which should be measured by kilograms.

Name_____

Learn You can measure how hot or cold something is. These thermometers show temperature in degrees Fahrenheit (°F).

Math Words
temperature
degrees

It is hot! It is 90°F.

It is cold! It is 20°F.

Try It Write the temperature.

1

_**40**_ °F

2

_____ °F

3 ✏️ Write **About It!** What does the red part on a thermometer show?

© Macmillan/McGraw-Hill

You can also measure temperature in degrees Celsius (°C).

0 °C is cold.

30 °C is hot.

Practice Write the temperature.

4 __10__ °C

5 _____ °C

6 _____ °C

7 _____ °C

Make it Right

THINK
SOLVE
EXPLAIN

8 Tim says the temperature is 65°F.

Tell what mistake he made. Make it right.

Math at Home: Your child read thermometers using Fahrenheit and Celsius scales.
Activity: Tell your child what the temperature was yesterday. Ask your child to estimate today's temperature in degrees Fahrenheit.

Problem Solving Strategy

Name_____

Use Logical Reasoning

You can use logical reasoning to find which measuring tool to use.

Dan wants to find how long his foot is. Which tool should he use to measure?

How heavy is it?   How long is it?   How hot is it?   How much does it hold?

Read

What do I already know? Dan wants to measure his _foot_.

What do I need to find? _____

Plan

I need to see which tool measures how long something is.

Solve

The  measures how _____.

Dan can use the  to see how long his foot is.

Look Back

Does my answer make sense?   Yes.   No.

© Macmillan/McGraw-Hill

Circle the tool to measure.

1 Susan wants to measure how long the ribbon is.

ribbon

2 John has a backpack. He wants to know how heavy it is.

backpack

3 Kate wants to know how much water her bucket would hold.

bucket

4 Jake wants to play outside. He wants to know how hot it is.

outside

5 Rosa wants to know how long her barrette is.

barrette

Problem Solving

Math at Home: Your child decided which measuring tool to use.
Activity: Ask your child which tool would be best for measuring the length of the kitchen, the amount a drinking glass holds, and the weight of an apple.

Name_____

The Measuring Track

▶ Put your counter on Start.

▶ Take turns. Toss the coin.

▶ Go 1 space for heads, 2 for tails.

▶ If the space tells how full, color a "How Full?" box on your score sheet. Do the same for "How Heavy?" and "How Hot?"

▶ The winner is the first to color all 6 boxes.

 2 players

 You Will Need

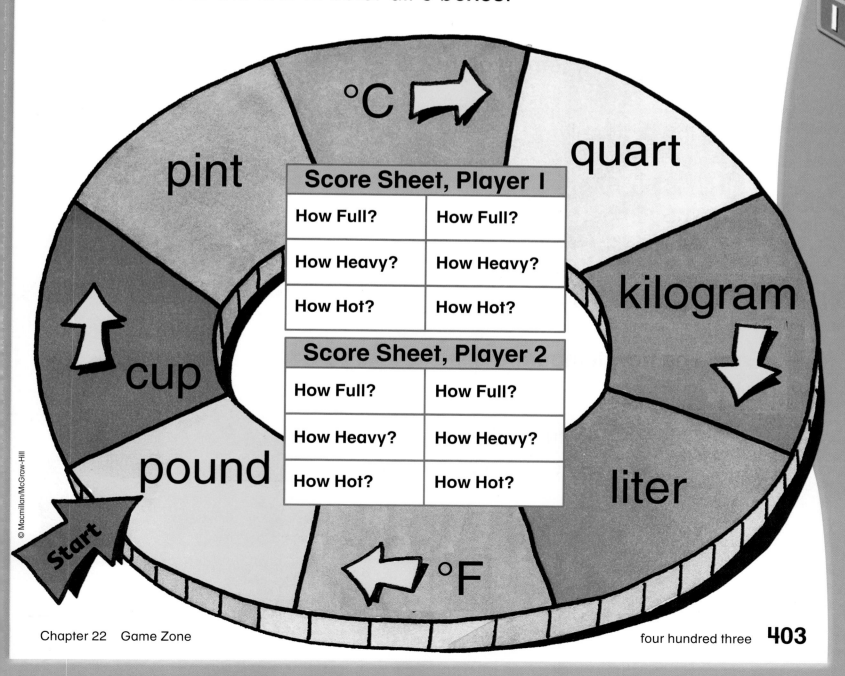

Score Sheet, Player 1	
How Full?	How Full?
How Heavy?	How Heavy?
How Hot?	How Hot?

Score Sheet, Player 2	
How Full?	How Full?
How Heavy?	How Heavy?
How Hot?	How Hot?

°C ⇨ quart kilogram liter °F pound cup pint

© Macmillan/McGraw-Hill

Technology Link

Temperature Changes • Calculator

You can use a to solve problems about temperature.

It is 60°F.
The temperature goes up 5°F.

Press

  6 **0** ⟶ **60**

+ **5** **=** ⟶ **65**

What is the new temperature? **65** °F

Use the  to solve.

1 It is 55°F.
The temperature goes down 10°F.
What is the new temperature?

Press (Clear) **5** **5** **−** **1** **0** **=** [ ]

The new temperature is _____ °F.

2 It is 75°F.
The temperature goes up 5°F.
What is the new temperature?

Press (Clear) **7** **5** **+** **5** **=** [ ]

The new temperature is _____ °F.

Name_____

Circle the best estimate.

1  less than 1 cup

more than 1 cup

2  less than 1 pint

more than 1 pint

3 less than 1 quart

more than 1 quart

4  less than 1 pound

more than 1 pound

Circle your answer.

5 Which container holds about 1 liter?

6 What should Ina use to weigh the ?

grams   kilograms

Circle the tool to measure.

7 Sue has a pencil. She wants to know how long it is.

8 Tim has a book. He wants to know how heavy it is.

Write the temperature.

9  _____ °F

10  _____ °C

© Macmillan/McGraw-Hill

Assessment

Choose the best answer.

1 You can trade 40 pennies for how many dimes?

⬭   ⬭   ⬭

2 Circle is greater than, is less than, or is equal to.

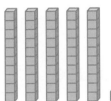 **51**

is greater than

is less than

is equal to

15

3 Count the tallies. Write how many.

卌 ||| _____

4 Count back to subtract.

$10 - 3 = $ _____

5 What number comes between 19 and 21? How do you know?

_____

_____

Exploring Shapes

READ TOGETHER

City Riddle

My cardboard city's being built,

So many shapes to see!

The purple squares each have 4 sides,

Green triangles have 3.

Red rectangles are fun to use,

For each side's not the same.

And there's a shape that has no sides—

Can you guess its name?

ANSWER: circle

Math at Home

Dear Family,

I will learn about 2- and 3-dimensional figures in Chapter 23. Here are my math words and an activity we can do together.

Love, _____

My Math Words

3-dimensional figures :

cube

rectangular prism

sphere

cylinder

cone

pyramid

2-dimensional shapes :

square

rectangle

circle

triangle

www.mmhmath.com
For Real World Math Activities

Home Activity

Ask your child to find about five boxes in your house. Have your child tell how the boxes are alike and different.

© Macmillan/McGraw-Hill

Books to Read

Look for these books at your local library and use them to help your child explore shapes.

- **The Wings on a Flea** by Ed Emberley, Little, Brown & Company, 2001.
- **The Shapes Game** by Paul Rogers, Henry Holt and Company, 1989.
- **Round Is a Pancake** by Joan Sullivan Baranski, Dutton Books, 2001.

Name_____

3-Dimensional Figures

Learn Here are some 3-dimensional figures.
Some 3-dimensional figures have flat faces.

cube

sphere

cone

pyramid

cylinder

face → ← edge
rectangular prism

Math Words

3-dimensional figure
cube
sphere
cone
pyramid
cylinder
rectangular prism
face
edge

Your Turn Find an object in your classroom that matches each figure. Draw the object.

 ①
cube

②
sphere

 ③
cone

④
cylinder

 ⑤
pyramid

⑥
rectangular prism

 ⑦ Write **About It!** How is a cube different from a sphere?

Practice   Color the blocks.

(8)

cube   sphere   cone   pyramid   cylinder   rectangular prism

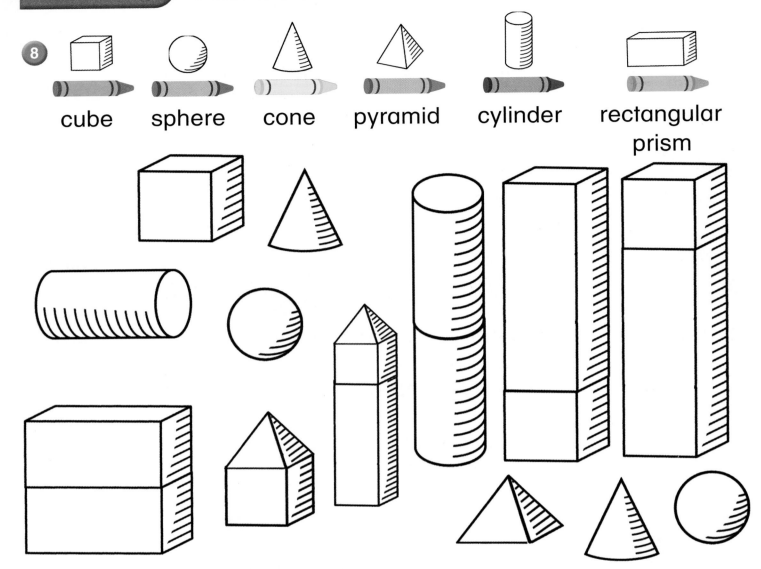

Problem Solving   Critical Thinking

THINK
SOLVE
EXPLAIN

(9) Sort the objects into two groups. Circle each object in one group. Underline each object in the other group.

Explain your sorting rule. _____

Math at Home: Your child identified 3-dimensional figures.
Activity: Have your child find objects shaped like cubes, spheres, cones, rectangular prisms, cylinders, and pyramids at home.

Name _____

Learn Trace around a flat face to make
a **2-dimensional shape** .

square

rectangle

circle

triangle

I traced around the
face of a cube and
made a square.

Math Words

**2-dimensional
shape**

square
rectangle
circle
triangle

Your Turn Find objects with shapes like these
in your classroom. Trace around one flat face.
Circle the shape you made.

 1

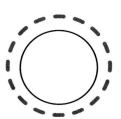

2

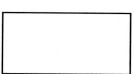

3

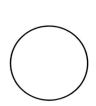

4
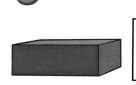
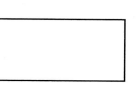

 5 ✏️ Write **About It!** Write how many faces are on the
object you used for Problem 4.

© Macmillan/McGraw-Hill

Practice Find objects with shapes like these in your classroom. Trace around all the flat faces. Write how many.

		☐ square	○ circle	△ triangle	▭ rectangle	Total number of flat faces
6		0	2	0	0	2
7						
8						
9						
10						

Problem Solving **Reasoning**

11 Color the 2-dimensional objects 🖍▶.

12 Color the 3-dimensional objects 🖍▶.

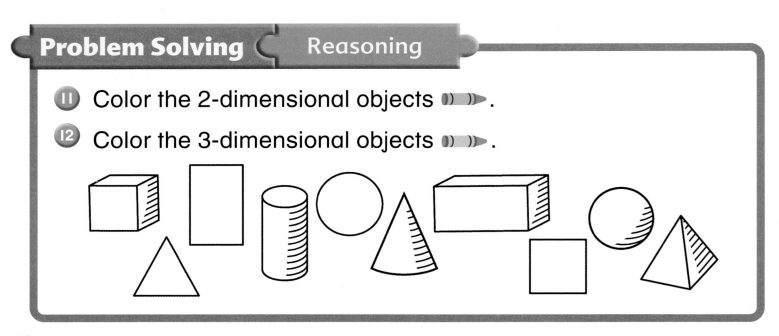

Math at Home: Your child traced around 3-dimensional figures to make circles, squares, rectangles, and triangles.
Activity: Show your child a can or a box and ask what shape he or she would make if he or she traced around it.

Name_____

Build Shapes

Learn You can build a new shape by putting other shapes together.

I made a trapezoid with three triangles.

Your Turn Use pattern blocks to make each shape. Draw how you made it.

Shape	Use	Draw Your Shape
1 hexagon	trapezoid	
2 hexagon	triangle	

3 ✏️ Write **About It!** Can you make a hexagon using a trapezoid and triangles? If yes, how?

© Macmillan/McGraw-Hill

Practice   Use pattern blocks to make each shape.
Draw how you made it.

Shape	Use	Draw Your Shape
4 trapezoid	parallelogram triangle	
5 parallelogram	triangle	
6 hexagon	trapezoid triangle	
7 hexagon	parallelogram triangle trapezoid	

Math at Home: Your child learned to build shapes by putting shapes together.
Activity: Ask your child what two shapes could be used to show a diamond shape.

Sides and Vertices

Learn Where sides meet is called a corner. Corners are also called vertices.

Math Words
side
corner
vertex
vertices
curve

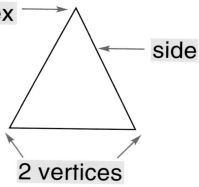

1 vertex ⟶

side ⟵

2 vertices

curve ⟶

A circle curves and has no sides or vertices.

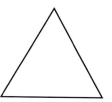

Try It Draw each shape.
Write how many sides and vertices.

1

___4___ sides

___4___ vertices

2

_____ sides

_____ vertices

3

_____ sides

_____ vertices

4

_____ sides

_____ vertices

5 ✏️ **Write About It!** Use sides and vertices to tell about a square.

© Macmillan/McGraw-Hill

Practice Look at the shapes.

All of these shapes have more than 3 sides and 3 vertices.

Color all the shapes that belong with each rule.

6 More than 3 sides

7 No sides and no vertices

8 3 vertices

9 4 sides and 4 vertices

Problem Solving **Visual Thinking**

10 I have 5 flat faces.
I have 8 edges.
What am I?

11 I have 6 flat faces.
I have 12 edges.
What am I?

Math at Home: Your child learned about sides and vertices of 2-dimensional shapes.
Activity: Draw a rectangle. Have your child tell how many sides and vertices it has.

Same Size and Same Shape

Learn These figures are the same size and same shape.

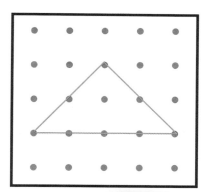 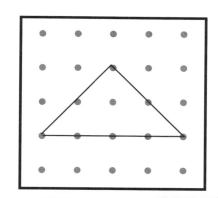

The triangles match.

Try It Draw a shape that matches.
Circle the name of the shape.

1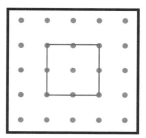

square
rectangle
triangle

2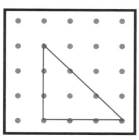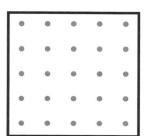

square
rectangle
triangle

3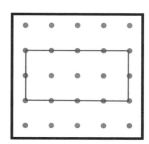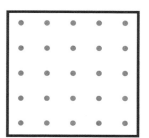

triangle
rectangle
circle

4 Write **About It!** Describe how a square is
different from a triangle.

© Macmillan/McGraw-Hill

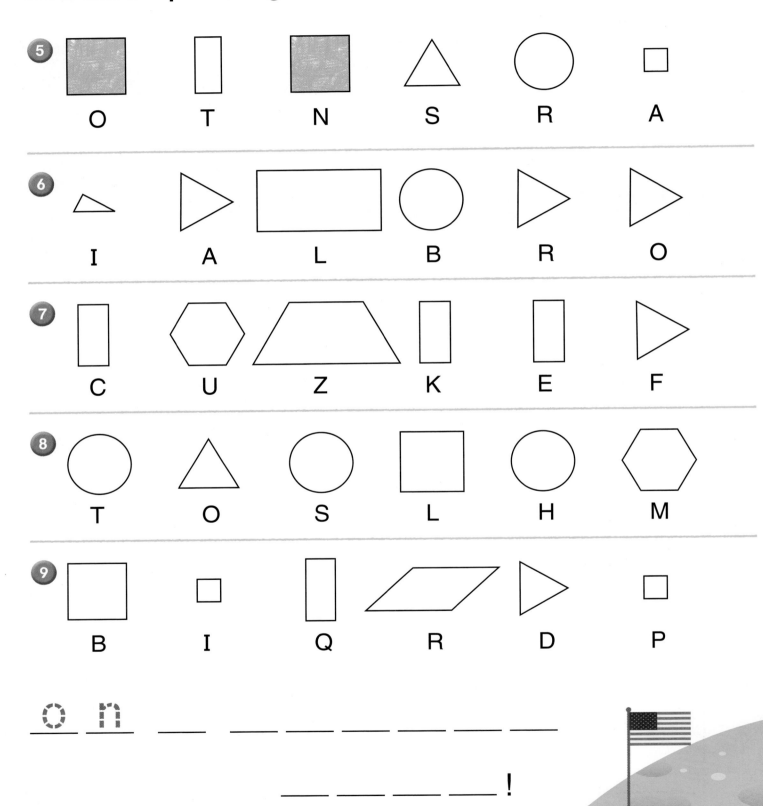

Practice
Color all the shapes that match.
Write the letters under the shapes
you colored in order at the bottom.

How did the pebbles get to the moon?

5   O   T   N   S   R   A

6   I   A   L   B   R   O

7   C   U   Z   K   E   F

8   T   O   S   L   H   M

9   B   I   Q   R   D   P

o n __ __ __ __ __ __ __ __

__ __ __ __ !

Math at Home: Your child showed how shapes can be alike or different.
Activity: Draw one large rectangle and one small rectangle.
Have your child tell how the two rectangles are alike and different.

Name_____

Solve.

Draw a line to the related subtraction fact.

$5 - 2 = 3$

$6 - 2 = 4$

$8 - 3 = 5$

$3 - 1 = 2$

$7 - 3 = 4$

$9 - 2 = 7$

$8 - 5 = $ ___

$5 - 3 = $ 2

$3 - 2 = $ ___

$6 - 4 = $ ___

$9 - 7 = $ ___

$7 - 4 = $ ___

© Macmillan/McGraw-Hill

Draw the next shape in each pattern.

1

2

3

4

5

6

7 ✏️ Write **About It!** Use claps and snaps to show the pattern for Problem 1.

LOG ON **www.mmhmath.com**
For more practice

Math at Home: Your child practiced finishing patterns.
Activity: Put some forks and spoons in a pattern. Ask your child to place the next utensil in the pattern.

420 four hundred twenty

Build a City

Jen and Rob make a block city.
They use 3-dimensional figures.
They use 2-dimensional shapes for
the signs on the trash can and truck.

Put
Trash
Here

J and R
Trucking

Problem Solving

Reading Skill   **Compare and Contrast**

1 How are the buildings different? _____

_____

Look at the signs on the trash can and truck.

2 How are the signs the same? _____

_____

3 How are the signs different? _____

_____

© Macmillan/McGraw-Hill

Build a Park

Jen and Rob build a park. They make places to sit and places to play. Look at all the different shapes in their park!

Problem Solving

Reading Skill **Compare and Contrast**

4 What figure was used to make the benches?

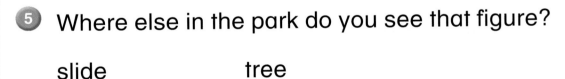

5 Where else in the park do you see that figure?

slide   tree

6 How are the two figures used to make a tree alike?

_____

7 How are they different?

_____

 Math at Home: Your child compared and contrasted information to answer questions.
Activity: Ask your child to choose two figures from the illustration, then tell how the figures are alike and different.

Problem Solving Practice

Solve.

1 Max uses a block to make a tree.
It has just one flat face. Circle the block.

2 Tina chooses the block with the most flat faces.
Which block does she choose? Circle the block.

 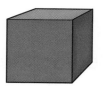 

THINK SOLVE EXPLAIN  Write a Story!

3 Jack uses 3-dimensional figures to make a house.
Write about the figures Jack might use.

_____

_____

_____

© Macmillan/McGraw-Hill

Writing for Math

THINK SOLVE EXPLAIN

Write about a way to sort these objects into two groups.

Writing

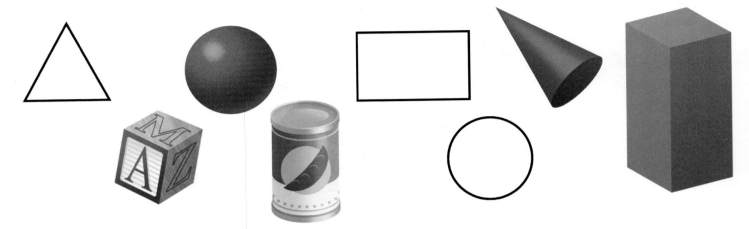

Think

How are some of these objects alike?

_____

Solve

I can sort now. I can underline the objects in one group and circle the objects in the other group.

Explain

I can tell how I sorted my groups.

_____

_____

e-Journal www.mmhmath.com
Write about math

Name_____

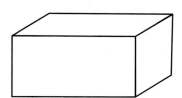

1 Color the 2-dimensional shapes .

2 Color the 3-dimensional figures .

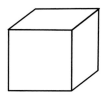  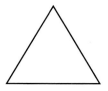

 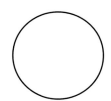 

Write how many sides and vertices.

3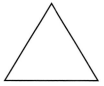

_____ sides

_____ vertices

4

_____ sides

_____ vertices

5 Sam picks a block with only 2 flat faces.
Which block does he choose? Circle the block.

 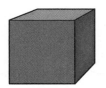 

© Macmillan/McGraw-Hill

Assessment

Spiral Review and Test Prep
Chapters 1—23

Choose the best answer.
Which is best for measuring the real object?

1  ◯ inch   ◯ foot   ◯ liter

2  ◯ inch   ◯ foot   ◯ liter

Write how many tens and ones.

3 **56** _____ tens _____ ones   **4** **80** _____ tens _____ ones

Write how many sides and vertices.

5  _____ sides

_____ vertices

6 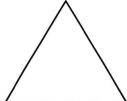 _____ sides

_____ vertices

THINK SOLVE EXPLAIN **7** Show two ways to make 45¢.

LOG⊘ON **www.mmhmath.com**
For more Review and Test Prep

Spatial Sense

Paper Fun

I had a piece of paper

And folded it in half,

And when I finished working,

I soon began to laugh.

For what I had discovered,

A real surprise to me,

The two parts matched exactly,

As you can clearly see!

Math at Home

Dear Family,

I will locate objects and learn about figures in Chapter 24. Here are my math words and an activity that we can do together.

Love, _____

My Math Words

slide:

F → F

turn:

F ↷ F

flip:

F ⟷ Ⅎ

Home Activity

Use a box. Choose an item. Use inside or outside as you ask your child where to place the item. Examples: "Put the ball outside the box." "Put the doll inside the box."

© Macmillan/McGraw-Hill

Books to Read

In addition to these library books, look for the **Time for Kids** math story that your child will bring home at the end of this unit.

- **Beep, Beep, Vroom, Vroom!** by Stuart J. Murphy, HarperCollins, 2000.
- **What's Next Nina?** by Sue Kassirer, The Kane Press, 2001.
- **Time for Kids**

LOG ON
www.mmhmath.com
For Real World Math Activities

Position

 Position words tell where objects are.

Math Words
over
under
in front of
behind
beside
to the right of
to the left of

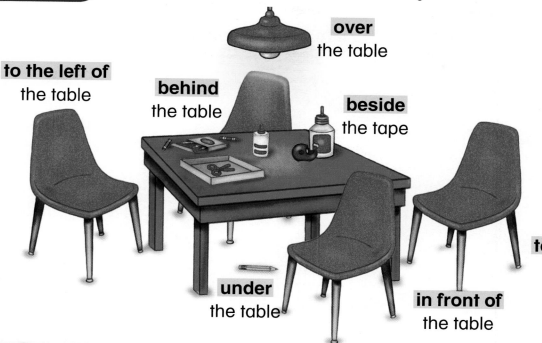

over
the table

to the left of
the table

behind
the table

beside
the tape

to the right of
the table

under
the table

in front of
the table

Try It Look at the picture. Circle the position word.

1 The crayons are _____ the scissors.

(behind)
to the left of

---

2 The tape is _____ the paste jar.

over

beside

---

3 The scissors is _____ the glue.

to the right of

to the left of

---

4 The table is _____ the light.

beside

under

---

5 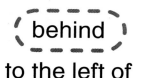 Write **About It!** Can the scissors have more than one position word to describe it? Explain.

© Macmillan/McGraw-Hill

Math Words

far
above
below
next to
down
up
near

far

above

below

down

up

next to

near

Draw.

6. 🧍 going up 🛝

7. 🧍 going down 🛝

8. 🌳 near 🪑

9. 🐦 above 🌳

10. 🐜 below 🪑

11. ⚽ next to 🪑

Math at Home: Your child used words to describe the position of objects.
Activity: Choose an object at home and have your child guess what it is by answering position clues such as: It is next to the television; it is below the window.

430 four hundred thirty

 **Learn** Some shapes are open and some are closed.

Math Words

open shape
closed shape

closed shape

open shape

Starts and ends at the same point.

Starts and ends at different points.

Try It Circle open or closed for each shape.

 1

Open
Closed

2

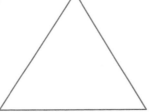

Open
Closed

3

Open
Closed

4

Open
Closed

5

Open
Closed

6

Open
Closed

7 Write **About It!** Is a square an open shape or a closed shape? Tell how you know.

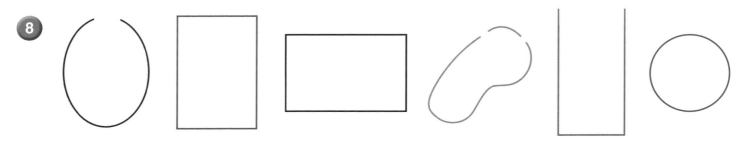

8

---

9 Draw 3 open shapes and 3 closed shapes.

10 Color the closed shapes.

Math at Home: Your child learned about open and closed shapes.
Activity: Have your child draw one open and one closed shape and explain the difference.

432 four hundred thirty-two

Name_____

Reflections of a Shape

Learn A reflection shows a shape as it would look in a mirror.

I made what the reflection shows.

Your Turn Use 📷 to make the shape and the reflection. Draw each reflection.

1

2

3

4

5 ✏ **Write About It!** How would your name look if you held it up to a mirror?

Practice
You may use .
Draw each reflection.

A reflection shows the shape backward.

6

7

8

9

10

11

Spiral Review and Test Prep

Choose the best answer.

12 What time is shown?

6:10   10:30   11:30
◯   ◯   ◯

13 What is the value of the coins?

41¢   51¢   56¢
◯   ◯   ◯

Math at Home: Your child learned about reflections of objects.
Activity: Show an object to your child and have him or her draw the reflection.
He or she can check by using a mirror.

Name_____

Slides, Turns, and Flips

Learn You can move objects with a slide, turn, or flip. Cut out the letter F.

Math Words
slide
turn
flip

slide | turn | flip

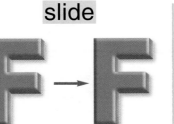 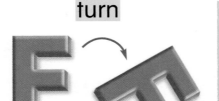 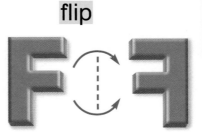

Use 1 finger to move it. | Use 2 fingers to twist it. | Pick it up and flip it to the back.

Try It Cut out the letters and move them as shown. Circle slide, flip, or turn to show how you moved them.

①

(slide)   turn   flip

②

slide   turn   flip

③

slide   turn   flip

④

slide   turn   flip

⑤ **Write About It!** What type of move will show you the back of an object if you are looking at the front?

© Macmillan/McGraw-Hill

You can cut out letters.
Circle slide, turn, or flip.

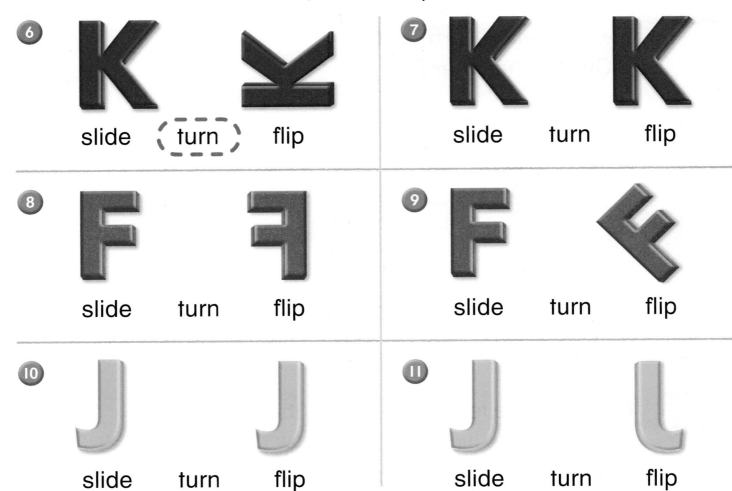

6 K ⅄

slide (turn) flip

7 K K

slide turn flip

8 F Ⅎ

slide turn flip

9 F ⌐

slide turn flip

10 J J

slide turn flip

11 J L

slide turn flip

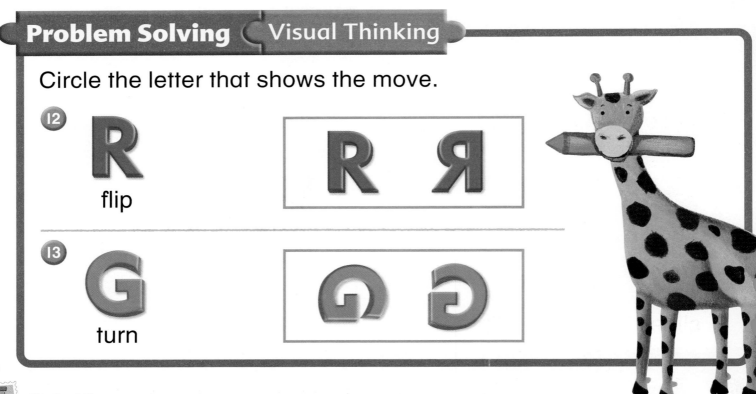

Problem Solving **Visual Thinking**

Circle the letter that shows the move.

12 R

flip

R Я

13 G

turn

ဌ Ə

 Math at Home: Your child learned about slides, turns, and flips.
Activity: Give your child a cut out letter E. Ask him or her to slide it.
Now have your child turn it and then flip it.

436 four hundred thirty-six

Name_____

Learn Shapes with symmetry have matching parts. A line of symmetry separates the matching parts.

Math Words
symmetry
line of symmetry

If you can fold a shape in half so it matches exactly, it has symmetry.

This shape has two equal parts.

This shape does not have equal parts.

Try It Circle the shapes with symmetry.

①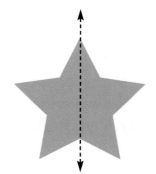

②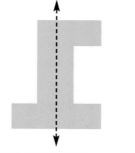

③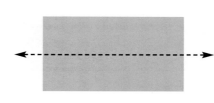

④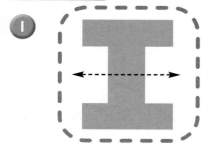

⑤

⑥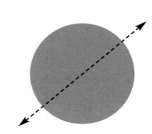

⑦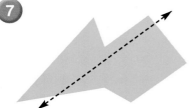

⑧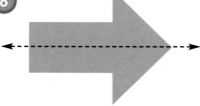

⑨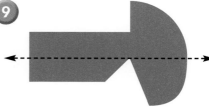

⑩ ✎ Write **About It!** Could a heart shape have symmetry? Why?

Practice Color the shapes with symmetry.

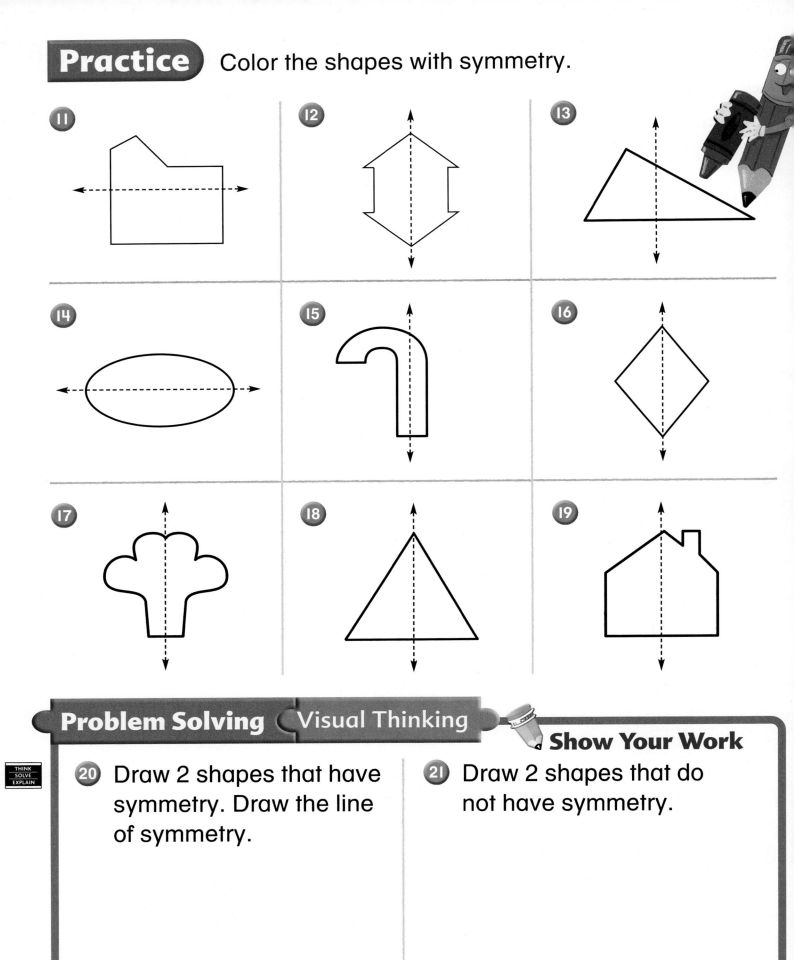

11

12

13

14

15

16

17

18

19

Problem Solving **Visual Thinking**

Show Your Work

THINK
SOLVE
EXPLAIN

20 Draw 2 shapes that have symmetry. Draw the line of symmetry.

21 Draw 2 shapes that do not have symmetry.

Math at Home: Your child learned to find shapes with symmetry.
Activity: Have your child find items in your house with symmetry.

438 four hundred thirty-eight

Problem Solving Strategy

Name_____

Describe Patterns ● Algebra

You can show a pattern in more than one way.

The pattern unit repeats over and over.
John uses A, B, and C to show the pattern
unit another way.
How will John's pattern look?

Math Word

pattern unit

pattern unit

A B C

Problem Solving

Read

What do I already know? _____

What do I need to find? _____

Plan

I need to repeat the pattern unit to complete John's pattern.

Solve

A B C ___ ___ ___ ___ ___ ___

Look Back

Does my answer make sense?   Yes.   No.

Circle the pattern unit.
Then use letters to show each pattern another way.

Problem Solving

1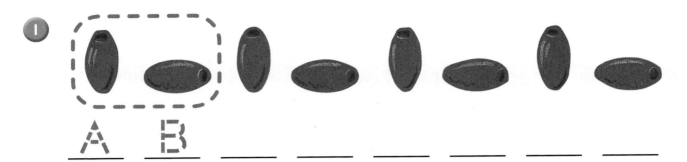

A B ___ ___ ___ ___

2

_____

3

_____

4

_____

Math at Home: Your child solved problems by describing a pattern in another way.
Activity: Have your child find patterns around the house, such as fabric or tile patterns.
Have your child tell the pattern unit and then describe the pattern in another way.

Name_____

Game Zone

Art Patterns

How to Play:

👥 2 players

▶ Pick 3 counters out of a bag. Make a pattern unit.

▶ Label the number cube 1, 1, 1, 2, 2, 2.

▶ Take turns. Toss the 🎲.

▶ Place that many pattern units on the path.

▶ The player who finishes the path wins.

You Will Need

🎲

12 ⚪

12 🔘

12 ⚫

Technology Link

Position of Shapes • Computer

- Use  to show position.

- Choose ◆.

- Stamp out ◆.

- Stamp out ◆.

- The ◆ is __under__ the ◆.

- The ◆ is _____ the ◆.

You can use the computer.
Tell the position.

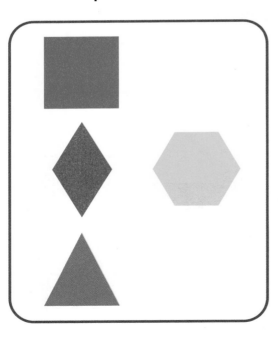

The ■ is _____ the ◆.

The ▲ is _____ the ◆.

The ◆ is _____ the ■ and the ▲.

The ⬡ is _____ the ◆.

 For more practice use Math Traveler.™

Name_____

Circle open or closed.

1 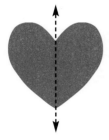 open

closed

2 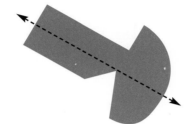 open

closed

Circle the shapes that show symmetry.

3

Circle the pattern unit. Then use letters to
show the pattern another way.

4

____ ____ ____ ____ ____ ____

5

____ ____ ____ ____ ____ ____ ____ ____

Assessment

Choose the best answer.

1 $8 + \blacksquare = 12$

⬭ 3   ⬭ 4   ⬭ 5

2 $12 - 7 = \blacksquare$

⬭ 3   ⬭ 4   ⬭ 5

3 Use tally marks to show seven.

What time is shown?

4

____:____

5

____:____

THINK SOLVE EXPLAIN **6** How can you measure the length of your desk without a ruler?

_____

_____

Name _____

It is fun to look for shapes on homes.

The door of this house is a rectangle.

One window is a circle.

Write how many of each shape you see.

□ _____

▷ _____

○ _____

▭ _____

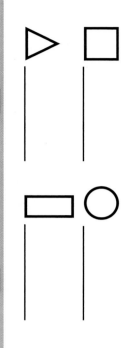

Fold down

Homes

People around the world live
in many kinds of homes.
Homes can have many
different shapes.

READ TOGETHER

© Macmillan/McGraw-Hill

Some homes look like rectangular prisms.

This home has both a cone shape and a cylinder shape.

Some homes have the shape of a cube.

Name _____

Use Shapes to Make Decisions

Plan to make a quilt with a pattern on it.
You can use ▰ ▱ ▲.

 You Decide!

1 Use the pattern blocks to make a pattern on the quilt.

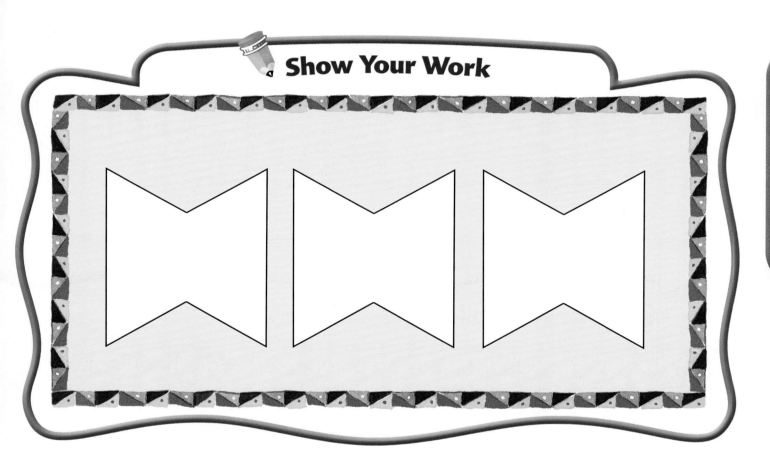

Show Your Work

2 Color your pattern.

 Your Decision!

3 Which pattern blocks did you use? Write how many.

_____ ▱   _____ ▰   _____ ▲

Make your own pattern. You can use 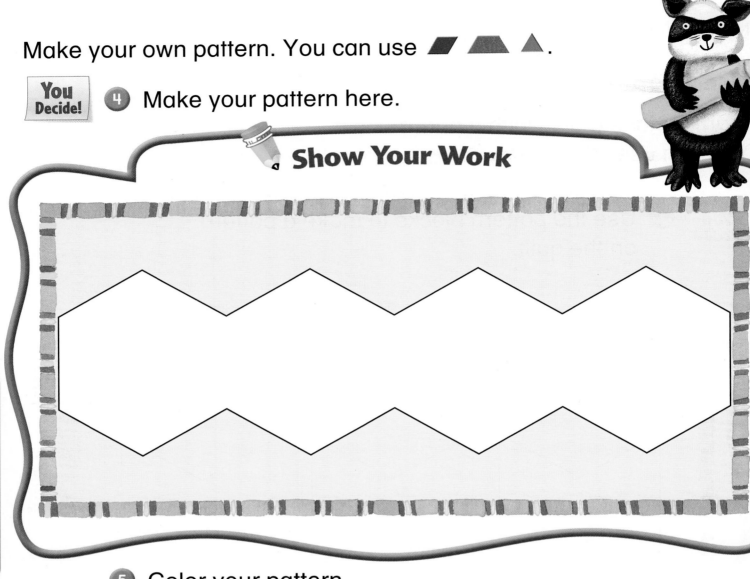.

You Decide! 4) Make your pattern here.

Show Your Work

5) Color your pattern.

Your Decision! 6) Which pattern blocks did you use? Write how many.

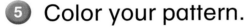

7) What if you used ■ and ▲ to make a pattern? What pattern could you make? Draw your pattern.

Math at Home: Your child used shapes to make decisions. Cut out paper shapes like those shown above.
Activity: Ask your child to combine shapes to make other symmetric shapes, such as two triangles to make a square.

Math Words

Draw lines to match.

1 Unit to measure length

2 Figure with no flat faces

3 Unit to measure weight

 sphere

pound

 inch

Skills and Applications

Measurement (pages 371–380, 389–400)

Examples

It is better to use feet to measure this real object.

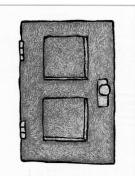

4  inches

feet

5  inches

feet

There is more than 1 liter of water in the pool.

6 more than 1 liter

less than 1 liter

7  more than 1 liter

less than 1 liter

Skills and Applications

Geometry (pages 409–420, 429–438)

Examples

 4 sides

4 vertices

8

 _____ sides

_____ vertices

The circled shape has symmetry.

9

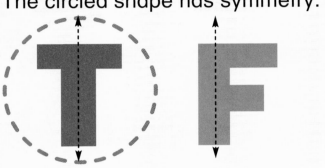

Problem Solving | **Strategy**

(pages 401–402, 439–440)

THINK SOLVE EXPLAIN

Circle the pattern unit. Then use letters to show the pattern another way.

10

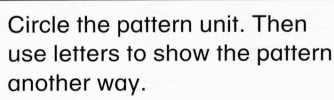

A B B A B B A B B

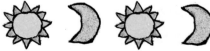

_____ _____ _____ _____ _____ _____

 Math at Home: Your child practiced measurement and geometry.
Activity: Have your child use these pages to review.

448 four hundred forty-eight

Unit 6 Review

Name _____

THINK
SOLVE
EXPLAIN

Unit 6
Performance Assessment

Draw Shapes

1 Find a small object in your classroom that is a rectangular prism.

- Draw a picture of it.

- Draw a circle above the picture.

- Draw a square below the picture.

2 Trace around a flat face of your object.

What shape did you make?

_____

Look at the shape. Write how many sides and vertices.

_____ sides _____ vertices

 You may want to put this page in your portfolio.

Assessment

Unit 6
Enrichment

Comparing Volume

Which holds more?

My glass holds 12 cubes.

My cup holds 7 cubes.

The glass holds more.

Find containers like these. Use  to fill each one.
Circle the one that holds more.

1

2

3

4

Fraction Concepts

READ TOGETHER

Class Picnic

Story by Gene McCormick • Illustrated by Nate Evans

Our class picnic really is a treat.
We share all the food that we eat.

How many equal parts do you see?

_____ equal parts

Here is more to share.
One is cut so it is fair.

Circle the one that shows equal parts.

How many of us can eat
this cherry pie so sweet?

How many equal parts? _____ equal parts

The pizza is so nice.
Each child takes a slice.
The pizza had equal parts.

How many equal parts did the pizza

have to start? _____ equal parts

Math at Home

Dear Family,

I will learn about equal parts and fractions in Chapter 25. Here are my math words and an activity that we can do together.

Love, _____

My Math Words

equal parts :
same size and shape

fraction :
part of a whole

   one half $\frac{1}{2}$:
I out of 2 equal parts

   one third $\frac{1}{3}$:
I out of 3 equal parts

   one fourth $\frac{1}{4}$:
I out of 4 equal parts

© Macmillan/McGraw-Hill

Home Activity

Have your child show ways to share food fairly with family or friends.

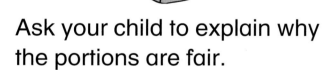

Ask your child to explain why the portions are fair.

Use foods such as oranges, bananas, sandwiches, pizza, and crackers.

Books to Read

Look for these books at your local library and use them to help your child learn about fractions.

- **Gator Pie** by Louise Matthews, Dodd, Mead, and Company, 1979.
- **Jump Kangaroo, Jump!** by Stuart J. Murphy, HarperCollins, 1999.
- **Rabbit and Hare Divide an Apple** by Harriet Ziefert, Viking Penguin, 1997.

Rabbit and Hare Divide an Apple
by Harriet Ziefert
Illustrated by Emily Bolam

LOG ON
www.mmhmath.com
For Real World Math Activities

Explore Fractions

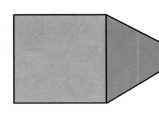

Learn You can use equal parts to make a whole.

This shape has 2 equal parts. The parts are the same size.

This shape has 2 unequal parts. The parts are not the same size.

Math Words
equal parts
unequal parts

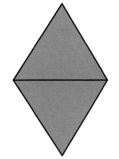

2 equal parts

2 unequal parts

Your Turn Place 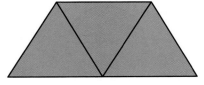 on the matching shapes. Write how many equal parts.

  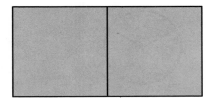

1.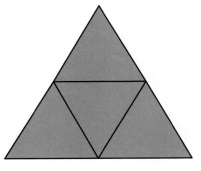

_3_ equal parts

2.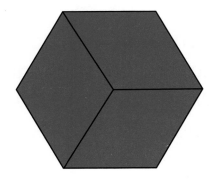

_____ equal parts

3.

_____ equal parts

4.

_____ equal parts

5. Write **About It!** How do you know if the parts are equal?

© Macmillan/McGraw-Hill

 equal parts

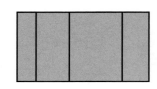 unequal parts

6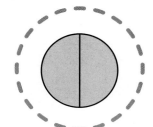

7

8

Problem Solving **Visual Thinking**

Show Your Work

THINK
SOLVE
EXPLAIN

9 4 children want to share the pie. Each child wants an equal part. Draw lines to show where you would cut the pie.

 Math at Home: Your child learned about equal and unequal parts.
Activity: Draw shapes divided into equal or unequal parts. Have your child identify them and explain how he or she knows which are equal and unequal.

454 four hundred fifty-four

Name_____

Math Words
fraction
one half
one third
one fourth

Learn You can use a 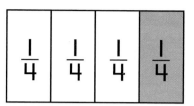fraction to tell about equal parts.

| $\frac{1}{2}$ | $\frac{1}{2}$ |

Halves
2 equal parts

1 out of 2 equal parts is blue.
One half is blue.

$\frac{1}{3}$ $\frac{1}{3}$ $\frac{1}{3}$

Thirds
3 equal parts

1 out of 3 equal parts is red.
One third is red.

$\frac{1}{4}$ $\frac{1}{4}$ $\frac{1}{4}$ $\frac{1}{4}$

Fourths
4 equal parts

1 out of 4 equal parts is green.
One fourth is green.

Try It Color one part ▬▬▶. Complete the sentence. Then write the fraction.

 1

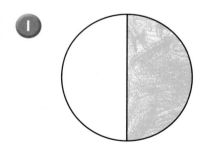

___1___ out of ___2___
equal parts is blue.

$\frac{1}{2}$ blue

2

_____ out of _____
equal parts is blue.

□
□ blue

 3 ✎ **Write About It!** A pizza is cut into 4 equal parts. 1 out of 4 equal parts has mushrooms. What fraction has mushrooms?

Color one part . Complete the sentence. Then write the fraction.

4

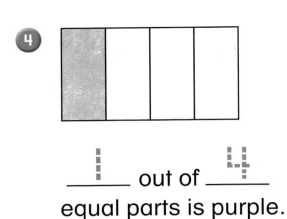

Each part is one fourth of the whole. I out of 4 equal parts is purple.

_____ out of _____ equal parts is purple.

$$\frac{1}{4}$$ purple

5

_____ out of _____ equal parts is purple.

_____ purple

6

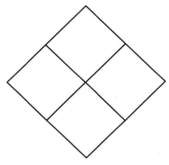

_____ out of _____ equal parts is purple.

_____ purple

Problem Solving Visual Thinking

7 Beth eats one part of a . How much does she eat? Circle your answer.

$$\frac{1}{2} \qquad \frac{1}{3} \qquad \frac{1}{4}$$

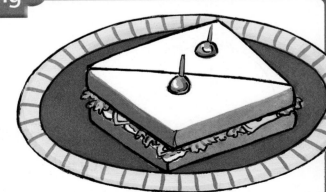

Math at Home: Your child identified unit fractions, such as 1/2, 1/3, and 1/4.
Activity: Fold pieces of paper to show halves and fourths. Have him or her color one part of each folded paper and tell what fraction is colored.

456   four hundred fifty-six

Name_____

Learn Some fractions name more than 1 equal part.

The top number tells how many blue parts. The bottom number tells how many parts in all.

$\frac{3}{4}$

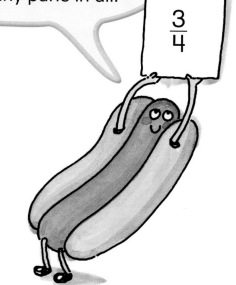

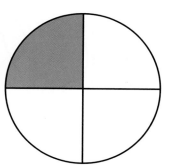   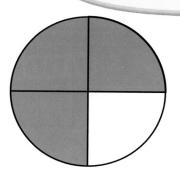

1 out of 4 equal parts
$\frac{1}{4}$ blue

2 out of 4 equal parts
$\frac{2}{4}$ blue

3 out of 4 equal parts
$\frac{3}{4}$ blue

Try It Color to show each fraction.

1 Color $\frac{1}{4}$.

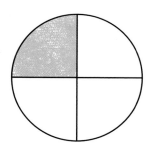

2 Color $\frac{2}{3}$.

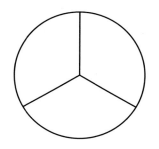

3 Color $\frac{4}{5}$.

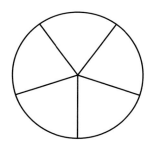

4 Color $\frac{2}{4}$.

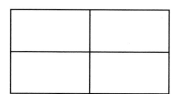

5 Color $\frac{2}{5}$.

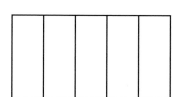

6 Color $\frac{6}{8}$.

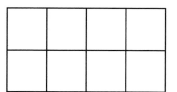

7   Write **About It!** How would you give a friend $\frac{2}{4}$ of a 🍞 ?

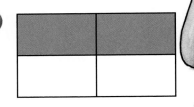

6 out of 8 equal parts
are green. $\frac{6}{8}$ green

8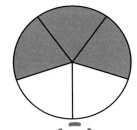

$\frac{1}{5}$ $\left(\frac{3}{5}\right)$ $\frac{5}{5}$

9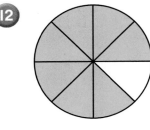

$\frac{1}{3}$ $\frac{2}{3}$ $\frac{3}{3}$

10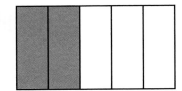

$\frac{1}{4}$ $\frac{2}{4}$ $\frac{3}{4}$

11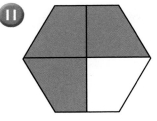

$\frac{1}{4}$ $\frac{2}{4}$ $\frac{3}{4}$

12

$\frac{1}{8}$ $\frac{3}{8}$ $\frac{7}{8}$

13

$\frac{2}{5}$ $\frac{3}{5}$ $\frac{5}{5}$

Problem Solving ⊂ Reasoning

THINK
SOLVE
EXPLAIN

Use the picture to solve.

14 What fraction of the pizza is left?

_____ out of _____ equal parts left.

$$\frac{\quad}{\quad}$$

of the pizza is left.

Math at Home: Your child learned about fractions that tell about more than one part, such as 2/3, 3/4, and 7/8.
Activity: Cut a food item into 4 equal parts. Have your child eat 1 part at a time and tell you how much is left (3/4, 2/3, 1/4).

458 four hundred fifty-eight

Name_____

Learn You can count parts to find a fraction for 1 whole.
This shape has 2 parts.

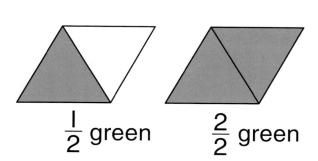

$\frac{1}{2}$ green $\frac{2}{2}$ green

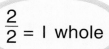
$\frac{2}{2}$ = 1 whole

The fraction for the whole equals 1.

Your Turn You may use pattern blocks.
Write the fraction for the whole.

1

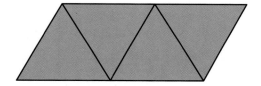

$\frac{4}{4}$ = 1 whole

2

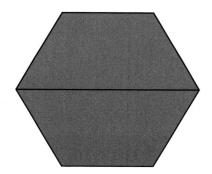

$\frac{}{}$ = 1 whole

3 Write **About It!** If you have $\frac{3}{3}$ of a pizza, how much of the pizza do you have?

You may use pattern blocks.
Write the fraction for the whole.

4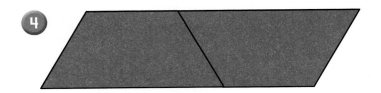

$$\frac{2}{2}$$ = I whole

5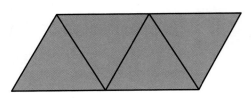

$$\frac{}{}$$ = I whole

6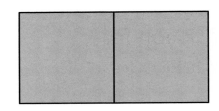

$$\frac{}{}$$ = I whole

7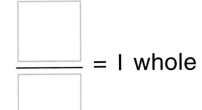

$$\frac{}{}$$ = I whole

Problem Solving **Critical Thinking**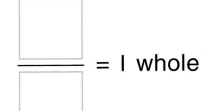

Show Your Work

Solve.

THINK
SOLVE
EXPLAIN

8 4 children eat lunch.
Each child has $\frac{1}{4}$ of an .
How many
do they have in all?

_____

 Math at Home: Your child identified fractions equal to 1 whole, such as 3/3 and 4/4.
Activity: Cut an orange, sandwich, or another food item into 4 or fewer equal parts. Have your
child count the parts and name the fraction for the whole.

Name_____

Write the time.

Add or subtract.

1. $6 - 2 =$

2. $7 + 8 =$

3. $10 - 0 =$

4. $4 - 1 =$

5. $8 + 6 =$

6. $8 - 4 =$

7. $6 + 5 =$

8. $9 - 4 =$

9. $8 + 2 =$

10. $9 + 9 =$

11. $4 - 3 =$

12. $7 + 5 =$

Math at Home: Your child practiced facts to 18.
Activity: Copy ten of these facts. Time your child as he or she writes the answers again.

Name_____

Fractions of a Group

Learn You can use a fraction to tell about parts of a group. This group has 4 picnic baskets. 1 of the 4 picnic baskets is yellow. So, $\frac{1}{4}$ of the group is yellow.

Try It Write how many are in the group. Circle 1 part of the group. Write the fraction that names the part you circled.

1.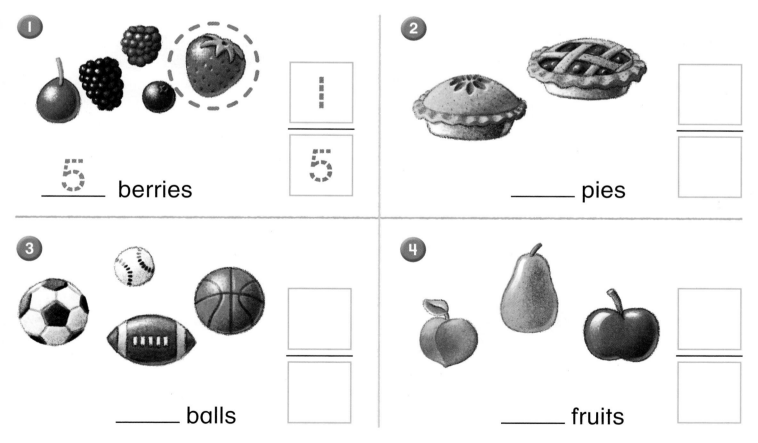

   __5__ berries $\dfrac{1}{5}$

2. _____ pies

3. _____ balls

4. _____ fruits

5.   Write **About It!** What does the bottom number of a fraction stand for?

Write how many are in the group.
Circle 1 part of the group. Write the
fraction that names the part you circled.

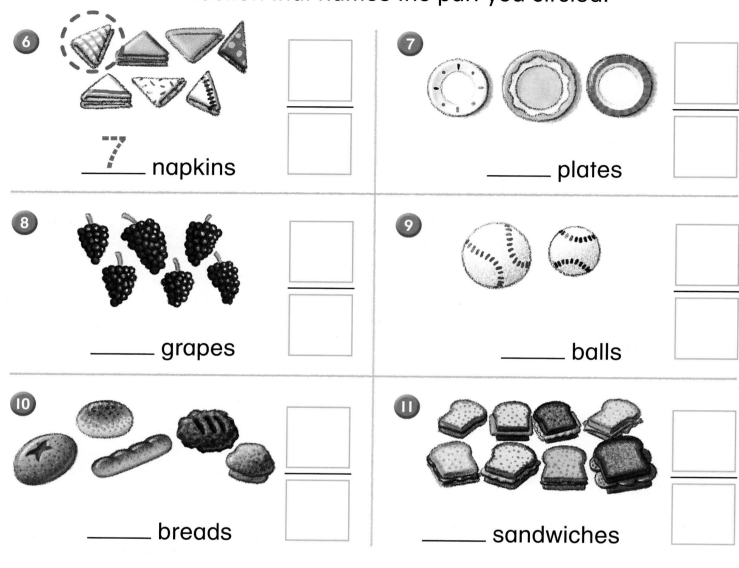

6 _____7_____ napkins

7 _____ plates

8 _____ grapes

9 _____ balls

10 _____ breads

11 _____ sandwiches

Make it Right

12 Tyrone wrote this fraction.

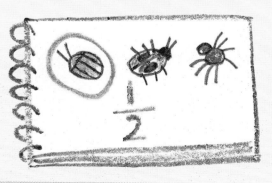

Why is Tyrone wrong?
Make it right.

_____

_____

Math at Home: Your child identified fractions of a group.
Activity: Show your child any four objects. For example, show 3 red beans and 1 white bean.
Have your child write a fraction to show the fraction that stands for 1 bean out of 4 beans.

464 four hundred sixty-four

Family Picnic

Family picnics are fun.
You play games outside.
You eat lots of good food.

Problem Solving

 Use Illustrations

1 The sandwich is cut into _____.

2 The pie is cut into _____.

3 Part of the group is sitting. Write the fraction. _____

4 Part of the apples are green. Write the fraction. _____

Picnic Time Is Over

Now, it is time to clean up.
There is some food left over.
Grandpa will take it home!

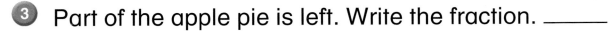

Reading Skill **Use Illustrations**

1. What part of the sandwich is left? _____

2. How many people are standing?

   _____ out of _____ people are standing.

3. Part of the apple pie is left. Write the fraction. _____

4. Part of the apples are green. Write the fraction. _____

Math at Home: Your child used illustrations to answer questions.
Activity: Ask your child to draw a picture that shows the fraction 2/3.

Problem Solving Practice

Solve.

---

1 4 people equally share a 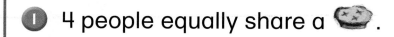.

What part does each person get? _____

If they eat 3 parts,
how much pie is left? _____

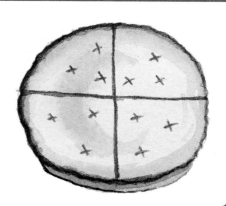

---

2 5 people will eat pizza.
If each person has 1 slice,
which part of the pizza
will they eat?   _____

Which part of the pizza
will be left?   _____

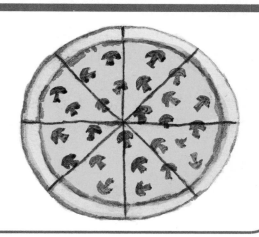

THINK
SOLVE
EXPLAIN   ✎ **Write a Story!**

3 Rita brought 3 yellow bowls
and 2 blue bowls to the picnic.
Which part of the group is blue? _____

Which part of the group is yellow? _____

Explain your answers.

_____

_____

Writing for Math

 THINK
SOLVE
EXPLAIN

Each mouse gets the same amount of cheese.

Write a fraction story about the mice and cheese.

Think

How many mice are there? _____

I need to cut 1 piece of cheese for each mouse.

Solve

I can write my fraction problem now.

_____

_____

_____

Explain

I can tell you why my problem and answer make sense.

_____

_____

_____

e-Journal www.mmhmath.com
Write about math

Name_____

1 Color one part 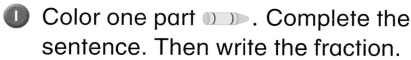. Complete the sentence. Then write the fraction.

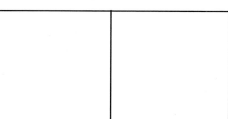

_____ out of _____ equal parts is yellow.

 yellow

2 Write the fraction for the whole.

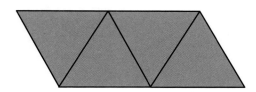

 = 1 whole

3 Color to show the fraction.

Color $\frac{2}{3}$.

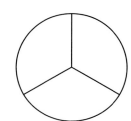

4 Write the fraction.

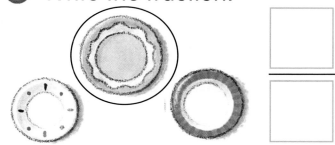

5 Write how many are in the group. Circle 1 part of the group. Write the fraction that names the part you circled.

_____ napkins

© Macmillan/McGraw-Hill

Assessment

Choose the best answer.

1 Which shape is a circle?

2 Which shape does not have a line of symmetry?

3 Draw the next 3 shapes in the pattern.

4 Write an addition fact that has the same sum as $8 + 7$.

____ ◯ ____ ◯ ____

5 Write the related addition facts for $16 - 7 = 9$.

____ ◯ ____ ◯ ____

____ ◯ ____ ◯ ____

THINK SOLVE EXPLAIN

6 Draw dots on the domino to show a double. Then write the doubles fact.

____ ◯ ____ ◯ ____

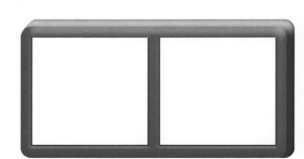

Fractions and Probability

Let's Eat!

Here's a pizza. What a treat!

Now it's time for us to eat.

But before our lunch can start,

Each one needs an equal part.

Math at Home

Dear Family,

I will learn about comparing fractions and probability in Chapter 26. Here are my math words and an activity that we can do together.

Love, _____

My Math Words

equally likely :

equally likely to spin ● or ●.

more likely , less likely :

more likely to spin ●.
less likely to spin ●.

impossible :

impossible to pick a blue marble

Home Activity

Cut food items, such as sandwiches or pizza, into halves, thirds, fourths, sixths, or eighths.

fourths
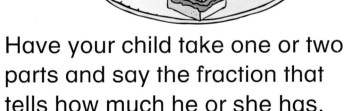

Have your child take one or two parts and say the fraction that tells how much he or she has.

Books to Read

Look for these books at your local library and use them to help your child learn about fractions and probability.

- **No Fair!** by Caren Holtzman, Scholastic, 1997.
- **How Many Ways Can You Cut a Pie?** by Jane Belk Moncure, Child's World, Inc., 1987.
- **Probably Pistachio** by Stuart J. Murphy, HarperCollins, 2001.

www.mmhmath.com
For Real World Math Activities

© Macmillan/McGraw-Hill

Name_____

Learn You can compare fractions.

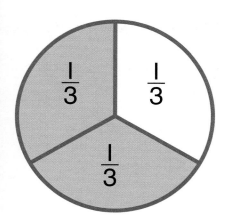

$\frac{1}{3}$ $\frac{1}{3}$ $\frac{1}{3}$

$\frac{2}{3}$ green

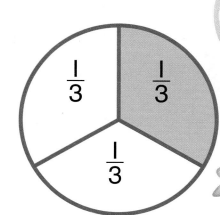

$\frac{1}{3}$ $\frac{1}{3}$ $\frac{1}{3}$

$\frac{1}{3}$ green

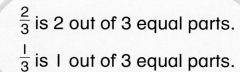

$\frac{2}{3}$ is 2 out of 3 equal parts.

$\frac{1}{3}$ is 1 out of 3 equal parts.

$\frac{2}{3}$ is greater than $\frac{1}{3}$

Try It Color to show each fraction.
Circle the greater fraction.

 1

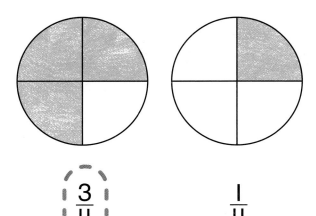

$\frac{3}{4}$ $\frac{1}{4}$

2

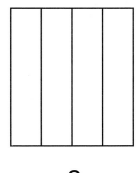

$\frac{3}{4}$ $\frac{1}{4}$

3  Write **About It!** $\frac{3}{4}$ of a sandwich is more than $\frac{2}{4}$ of a sandwich.
Draw a picture to show how you know.

Color the food to show each fraction.
Circle the greater fraction.

Compare. Which is greater?

4  $\frac{1}{8}$  $\frac{2}{8}$

5 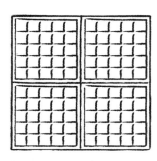 $\frac{1}{4}$ 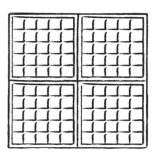 $\frac{3}{4}$

6  $\frac{4}{5}$  $\frac{2}{5}$

Problem Solving   Critical Thinking

THINK
SOLVE
EXPLAIN

Use the picture to solve.

7 $\frac{2}{5}$ are .

Write the
fraction for .

8 Write both fractions.
Circle the greater fraction.

 Math at Home: You child learned how to compare fractions.
Activity: Fold identical sheets of scrap paper into halves and fourths. Have your child cut out 1/2, 3/4, 1/4, and 2/4
of the paper and write the fraction on each cutout. Then have him or her use the cutouts to compare the fractions.

Name_____

Compare Unit Fractions

Learn You can compare unit fractions to see which is greater.

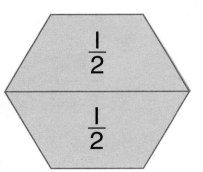

$\frac{1}{2}$
$\frac{1}{2}$

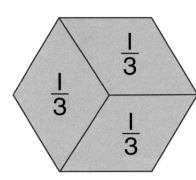

$\frac{1}{3}$
$\frac{1}{3}$
$\frac{1}{3}$

Each is $\frac{1}{2}$.　　Each ◢ is $\frac{1}{3}$.

A ⬡ can be covered with 3 ◢, but it only takes 2 ▲. So I know $\frac{1}{2}$ is greater than $\frac{1}{3}$.

Your Turn Use ◢▼▲ to make each shape. Compare 1 part of each shape. Write the fractions.

1

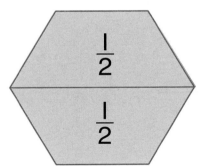

$\frac{1}{2}$
$\frac{1}{2}$

$\frac{1}{6}$ $\frac{1}{6}$ $\frac{1}{6}$
$\frac{1}{6}$ $\frac{1}{6}$ $\frac{1}{6}$

$\frac{1}{2}$ is greater than $\frac{1}{6}$

2

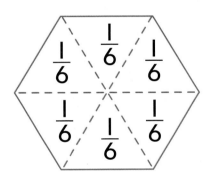

$\frac{1}{6}$ $\frac{1}{6}$ $\frac{1}{6}$
$\frac{1}{6}$ $\frac{1}{6}$ $\frac{1}{6}$

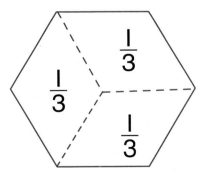

$\frac{1}{3}$
$\frac{1}{3}$
$\frac{1}{3}$

☐ is greater than ☐

3 ✎ **Write About It!** Which is larger, $\frac{1}{2}$ or $\frac{1}{4}$ of the same size circle? Draw to show your answer.

© Macmillan/McGraw-Hill

Practice Color 1 part to show each fraction.
Compare the parts you colored.
Circle the fraction that is greater.

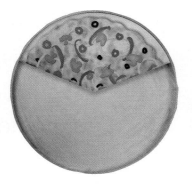

Compare to see which part is larger.

4 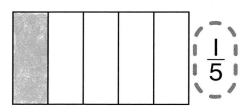 $\frac{1}{5}$ 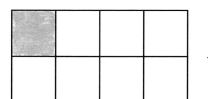 $\frac{1}{8}$

5  $\frac{1}{3}$  $\frac{1}{4}$

6 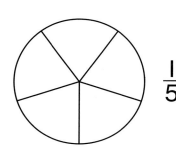 $\frac{1}{5}$ 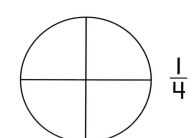 $\frac{1}{4}$

Problem Solving | Reasoning

Use the picture to solve.

Kim   Ben

7 Kim eats $\frac{1}{4}$ of a pizza.

Ben eats $\frac{1}{3}$ of a pizza.

Who eats more?
Circle the pizza slice.

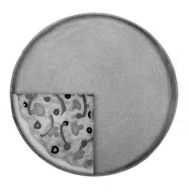

Math at Home: Your child compared unit fractions.
Activity: Cut fruits such as apples, oranges, or bananas, into halves, fourths, and eighths. Have your child use the pieces to compare 1/2 to 1/4, 1/2 to 1/8, and 1/4 to 1/8.

476 four hundred seventy-six

Equally Likely

Learn The spinner shows two possible outcomes.

Math Words

possible outcome
equally likely

The blue and orange parts are the same size.

The  on the spinner can land on ◯ or ●. It is equally likely to land on either color.

It is a possible outcome that the  will land on ◯ or ●.

Try It Circle the spinners that show it is equally likely that the  will land on ◯ or ●.

1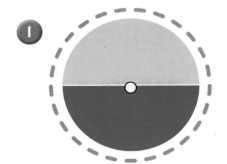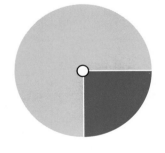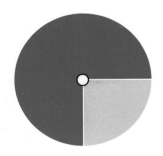

2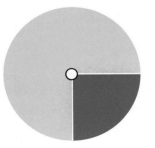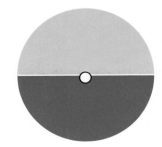

3 ✏️ Write **About It!** A spinner is $\frac{1}{2}$ blue and $\frac{1}{2}$ red. Is it equally likely to land on either color? Explain.

© Macmillan/McGraw-Hill

Practice   Some spinners are equally likely to land on either color. Circle them.

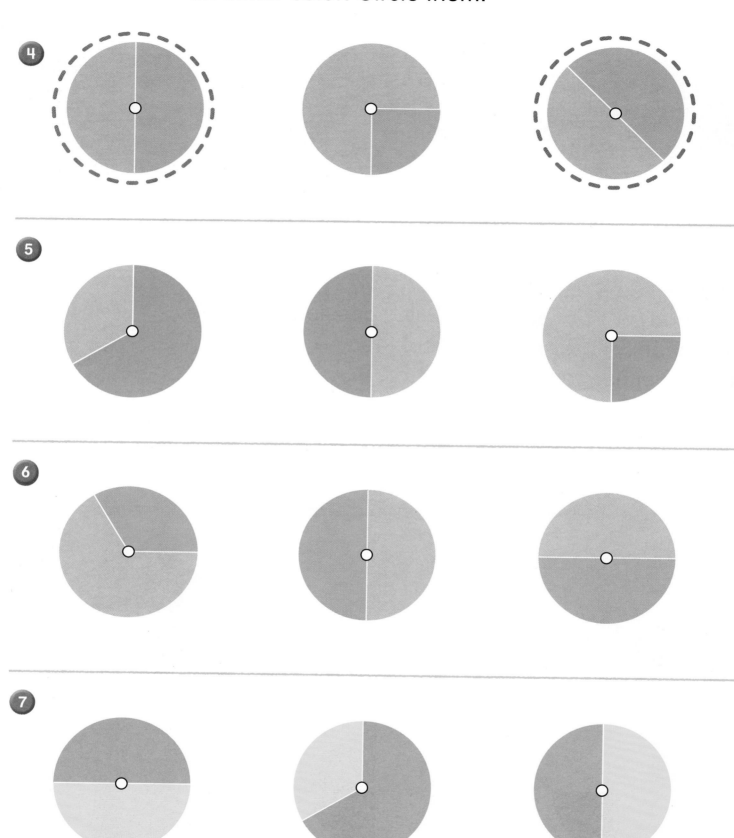

Math at Home: Your child learned about equally likely outcomes.
Activity: Have your child draw a spinner to show that it is equally likely to land on ● or ●.

478   four hundred seventy-eight

Name_____

More Likely and Less Likely

Learn You can predict if an outcome is more likely or less likely to happen. It is more likely you will pick a ▪.

> There are more ▪ than ▫.
> So, I am more likely to pick a ▪.
> It is less likely that I will pick a ▫.

Math Words
predict
more likely
less likely

Your Turn

- Put cubes in a paper bag as shown.

- Predict which color you are more likely to pick.

- Which color are you less likely to pick?

- Pick one cube without looking.

- Color each cube.

Bag	More Likely	Less Likely	Your Pick
1			
2			
3			

4 ✏ Write **About It!** There are 9 and 4 ▫. in a bag. Are you more likely to pick a  or ▫? Explain.

Practice

- Put cubes in a paper bag.
- Which color are you more likely to pick?
- Which color are you less likely to pick?
- Pick one cube without looking.
- Color each cube.

Bag	More Likely	Less Likely	Your Pick
5			
6			
7			
8			

Problem Solving — Reasoning

Circle more likely or less likely.

9 Skiing in Florida

    more likely   less likely

10 Swimming in Florida

    more likely   less likely

Math at Home: Your child learned about predicting outcomes and deciding whether an event is more likely, equally likely, or less likely to happen.
Activity: Put 1 penny and 5 nickels in a bag. Ask your child if he or she is more likely or less likely to pick a penny.

Name_____

Certain, Probable, Impossible

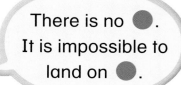

Learn You can predict if an outcome is
certain, probable, or impossible.

There is no 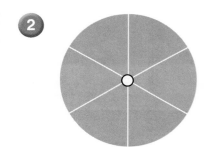.
It is impossible to
land on 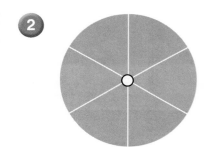.

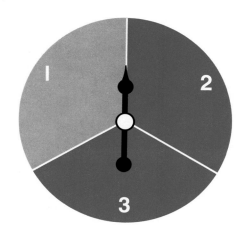

Math Words
certain
probable
impossible

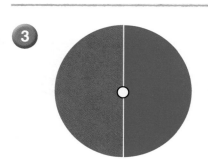

It is certain that the ⌁ will land on a number.
It is probable that the ⌁ will land on ●.
It is impossible that the ⌁ will land on ●.

Your Turn Use ⊕. Circle if it is certain, probable,
or impossible that the ⌁ will land on ●.

1️⃣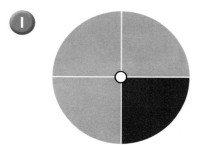

certain

(probable)

impossible

2️⃣

certain

probable

impossible

3️⃣

certain

probable

impossible

4️⃣

certain

probable

impossible

5️⃣ ✏️ Write **About It!** You reach into a jar of , , and . Is it
certain that you will pull out a ? Explain.

Use ⊕. Color the spinner so that it is certain, probable, or impossible that the the ⊶ will land on ●.

6 probable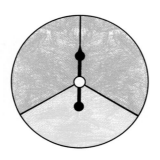

7 probable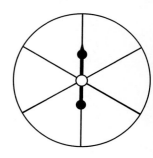

8 impossible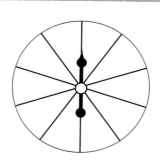

9 certain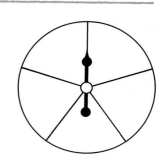

10 probable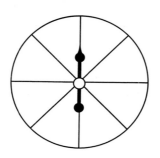

11 impossible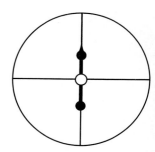

Spiral Review and Test Prep

Choose the best answer.

12 Which shape comes next?

○ ○ ○

13 Which fraction names the larger part?

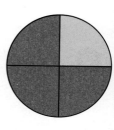

$\dfrac{1}{4}$ $\dfrac{2}{4}$ $\dfrac{3}{4}$

○   ○   ○

 Math at Home: Your child learned about probability and decided whether an event was certain, probable, or impossible.
Activity: Have your child name one event that is certain, such as the sun rising, one event that is probable, such as catching a ball, and one event that is impossible, such as a pencil talking.

Name_____

Problem Solving Strategy

Draw a Picture

You can draw a picture to solve problems.

Tomás puts 3 , 7 , and 2 ◯ in a bag. Which marble color is he most likely to pull out of the bag?

Read

What do I already know? _____ ● _____ ● _____ ◯

What do I need to find? _____

_____

Plan

I can draw a picture on the bag to help.

Solve

My picture shows that Tomás is most likely to pull out   ◯

Look Back

Does my answer make sense?   Yes.   No.

How do I know? _____

_____

Problem Solving

Draw a picture to solve.

Circle your answer.

1. Jill has 3 apples and 3 oranges in a bag. Is Jill more likely, equally likely, or less likely to pull out an apple?

   more likely

   (equally likely)

   less likely

2. Kelly has 1 raisin and 3 peanuts in a bag. Is it certain, probable, or impossible that she will pull out a grape?

   certain

   probable

   impossible

3. Sam has 1 blueberry, 6 cherries, and 1 grape in a bag. Is it certain, probable, or impossible that he will pull out a cherry?

   certain

   probable

   impossible

Problem Solving

Draw or write to explain.

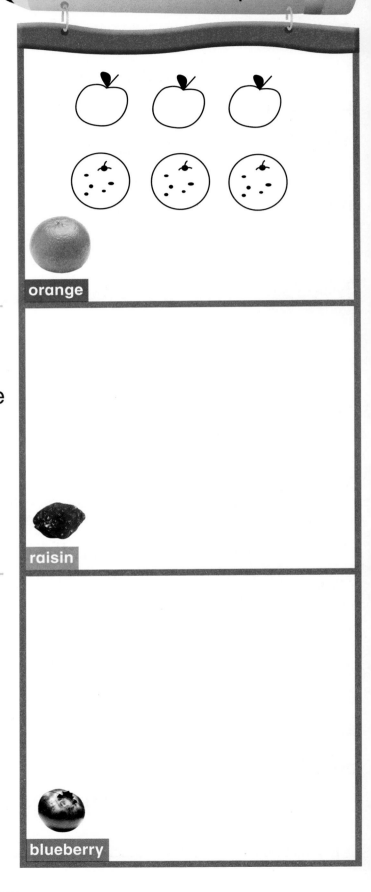

orange

raisin

blueberry

Math at Home: Your child solved probability problems by drawing pictures.
Activity: Ask your child the following problem: "I have 10 nickels and 4 quarters in my pocket. Which coin am I more likely to pull out?" Have your child draw pictures to solve.

Name_____

Predict Your Color

How to Play:

▶ Each player chooses one slice.

▶ Now pick a counter from the bag.

▶ If the color matches, put it on your slice.

▶ If not, put it back in the bag.

▶ The first player to fill his or her slice wins.

 2 players

You Will Need

7 ●

7 ●

Technology Link

Model Fractions • Computer

- Use 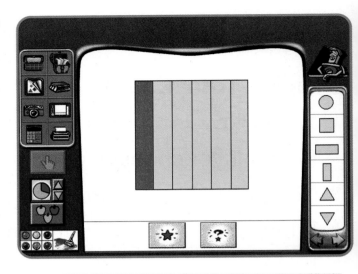.
- Stamp out a square.
- Click on the square.
- Click the up arrow 5 times.
- Now there are 6 equal parts.
- Color 1 part.
- The fraction $\frac{1}{6}$ names that part.

Show more fractions.
You can use the computer.

1 Make 3 equal parts.
Color 2 parts red.
Write the fraction.

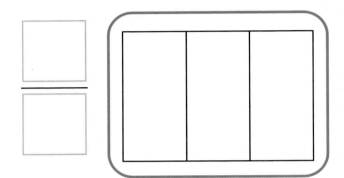

---

2 Make 4 equal parts.
Color 3 parts red.
Write the fraction.

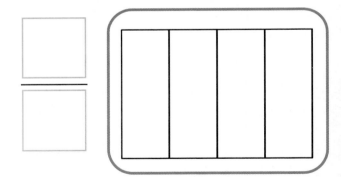

---

3 Make 8 equal parts.
Color 3 parts red.
Write the fraction.

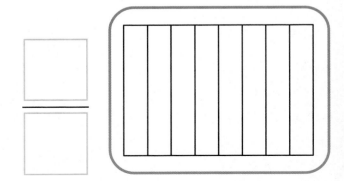

 For more practice use Math Traveler.™

Name_____

Color to show each fraction. Circle the greater fraction.

1 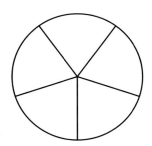 $\dfrac{3}{5}$ 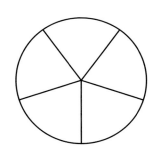 $\dfrac{2}{5}$

Circle the spinner that is equally likely to land on either color.

2

3

4 Circle if it is certain, probable, or impossible that the ⊶ will land on ●.

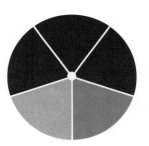

certain

probable

impossible

Draw a picture to solve. Circle your answer.

5 Pam has 1 🔲, 2 🔲, and 5 🔲 in a bag.
Is it more likely, equally likely, or less likely to pull a 🔲 from the bag?

more likely

equally likely

less likely

© Macmillan/McGraw-Hill

Spiral Review and Test Prep
Chapters 1—26

Choose the best answer.

1 I am greater than 21.
I am less than 36.
What number am I?

32 ◯   40 ◯   54 ◯

2 Rosa has 4 nickels. She buys a ring for 10¢. How much money does she have left?

30¢ ◯   20¢ ◯   10¢ ◯

3 Which weighs more than 1 pound?

a leaf ◯   a rabbit ◯   a sock ◯

Test Prep

4 Use the ruler to measure a .
Draw a line to show the length.

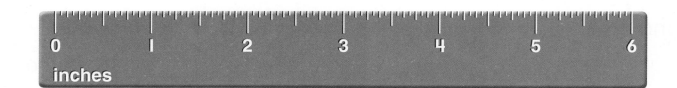

About how long is the crayon?   About _____ inches

 5 Show different ways to make 17¢.

Addition and Subtraction to 20

Pine Cone Treats

20 squirrels in a tree,

Munching pine cones carefully.

10 stop eating, scoot away.

10 are left, to snack all day!

Math at Home

Dear Family,

I will review facts to 20 and learn to use patterns to add and subtract in Chapter 27. Here are my math words and an activity that we can do together.

Love, _____

My Math Words

related facts:

$9 + 6 = 15$
$15 - 6 = 9$ $15 - 9 = 6$

fact family:

$7 + 5 = 12$
$5 + 7 = 12$
$12 - 5 = 7$
$12 - 7 = 5$

Home Activity

Cut an egg carton like this.

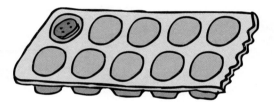

Place 1 button in the carton. Have your child find how many more are needed to fill the carton, and write an addition sentence.
$1 + 9 = 10$

© Macmillan/McGraw-Hill

Books to Read

Look for these books at your local library and use them to help your child with addition and subtraction.

- **Let's Count It Out, Jesse Bear** by Nancy White Carlstrom, Aladdin Paperbacks, 1996.
- **Safari Park** by Stuart J. Murphy, HarperCollins, 2002.

LOG ON
www.mmhmath.com
For Real World Math Activities

Name_____ **Patterns with 10**

Learn Number patterns can help you add.
Show 10. Show 3 more. What is the sum?

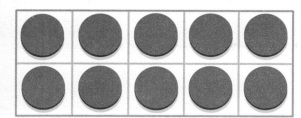

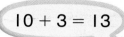
10 + 3 = 13

10 + 3 = 13

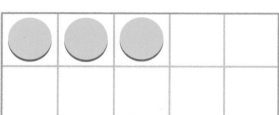

Your Turn Add. Look for a pattern. Use ⬤ and ▭▭▭▭ .

① 10 + 1 = 11	⑥ 10 + 6 = ___
② 10 + 2 = ___	⑦ 10 + 7 = ___
③ 10 + 3 = ___	⑧ 10 + 8 = ___
④ 10 + 4 = ___	⑨ 10 + 9 = ___
⑤ 10 + 5 = ___	⑩ 10 + 10 = ___

⑪ **Write About It!** Look at Exercises 1– 5.
What pattern do you see?

Show 10. Show 2 more. The sum is 12.

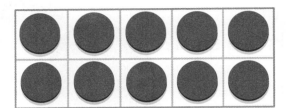 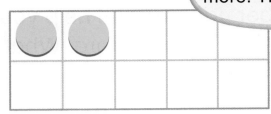

12 $10 + 2 = \underline{12}$

---

13 $10 + 4 = \underline{}$ **14** $10 + 7 = \underline{}$ **15** $10 + 5 = \underline{}$

16
$\begin{array}{r} 10 \\ +\ 1 \\ \hline \end{array}$
17
$\begin{array}{r} 10 \\ +\ 6 \\ \hline \end{array}$
18
$\begin{array}{r} 10 \\ +\ 9 \\ \hline \end{array}$
19
$\begin{array}{r} 10 \\ +\ 7 \\ \hline \end{array}$
20
$\begin{array}{r} 10 \\ +\ 3 \\ \hline \end{array}$
21
$\begin{array}{r} 10 \\ +\ 2 \\ \hline \end{array}$

22
$\begin{array}{r} 5 \\ +10 \\ \hline \end{array}$
23
$\begin{array}{r} 3 \\ +10 \\ \hline \end{array}$
24
$\begin{array}{r} 4 \\ +10 \\ \hline \end{array}$
25
$\begin{array}{r} 8 \\ +10 \\ \hline \end{array}$
26
$\begin{array}{r} 6 \\ +10 \\ \hline \end{array}$
27
$\begin{array}{r} 10 \\ +10 \\ \hline \end{array}$

Problem Solving **Number Sense**

Circle the best answer.

28 Which is another way to show 18?

29 Katie puts 14 in two rows. She puts 10 in the first row. How many does she put in the second row?

$8 + 1$ $10 + 8$ $10 + 1$

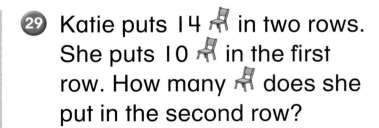
3 🪑   4 🪑   5 🪑

Math at Home: Your child learned how to add number patterns with 10.
Activity: Put 10 pennies in a row. Ask your child to add 1, 2, and 3 pennies to 10 and find the sum each time.

Name_____

Make a 10 to Add

 **Activity**

Learn Make a 10 to help you add.
Add 8 + 5.
Start with 8 ●. Then add 5 ○.

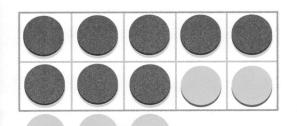

I used 2 ○ to make 10.
Then I added the other 3.
10 + 3 = 13

$10 + 3 = \underline{13}$

so

$8 + 5 = \underline{13}$

Your Turn Add. Use ○ and ▦.

① $9 + 5 = \underline{14}$ ② $8 + 6 = \underline{\hphantom{00}}$ ③ $7 + 5 = \underline{\hphantom{00}}$

④ $8 + 7 = \underline{\hphantom{00}}$ ⑤ $9 + 7 = \underline{\hphantom{00}}$ ⑥ $9 + 6 = \underline{\hphantom{00}}$

⑦ $\begin{array}{r} 9 \\ +4 \\ \hline \end{array}$ ⑧ $\begin{array}{r} 8 \\ +4 \\ \hline \end{array}$ ⑨ $\begin{array}{r} 7 \\ +8 \\ \hline \end{array}$ ⑩ $\begin{array}{r} 9 \\ +8 \\ \hline \end{array}$ ⑪ $\begin{array}{r} 7 \\ +6 \\ \hline \end{array}$ ⑫ $\begin{array}{r} 7 \\ +4 \\ \hline \end{array}$

⑬ Write **About It!** How can making a 10 help you find
the sum of 9 + 8?

Chapter 27   Lesson 2   four hundred ninety-three   **493**

Practice Add. Use 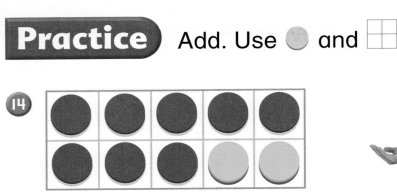 and ⊞.

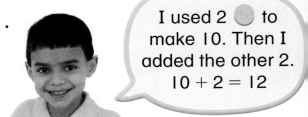
I used 2 ⬤ to make 10. Then I added the other 2. $10 + 2 = 12$

14

$8 + 4 = \underline{12}$

15 $7 + 4 = \underline{}$ 16 $8 + 7 = \underline{}$ 17 $9 + 4 = \underline{}$

18 $9 + 5 = \underline{}$ 19 $7 + 9 = \underline{}$ 20 $8 + 9 = \underline{}$

21 9 $+6$ 22 7 $+8$ 23 8 $+8$ 24 6 $+8$ 25 8 $+5$ 26 9 $+4$

27 9 $+8$ 28 7 $+6$ 29 7 $+7$ 30 8 $+3$ 31 9 $+3$ 32 5 $+9$

Problem Solving — Reasoning — **Show Your Work**

 THINK SOLVE EXPLAIN

Solve. Use ⬤ and ⊞.

33 $9 + 7$ is the same as $10 + \underline{}$.

Math at Home: Your child learned how to make a 10 when adding numbers.
Activity: Have your child use pennies to show you how 8 + 4 and 10 + 2 are the same.

Relate Addition and Subtraction

ALGEBRA
x

Learn Knowing related facts can help you add and subtract.

$$9 + 4 = 13$$

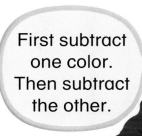

First subtract one color. Then subtract the other.

$$13 - 4 = 9$$
$$13 - 9 = 4$$

Try It Use ■ and □. Add. Then subtract. Write the related subtraction facts.

	Add	Subtract	Related subtraction facts
1	$5 + 8 = \underline{13}$	8	$\underline{13} \ominus \underline{8} = \underline{5}$
		5	$\underline{13} \ominus \underline{5} = \underline{8}$
2	$7 + 9 = \underline{\ \ }$	9	$\underline{\ \ } \bigcirc \underline{\ \ } = \underline{\ \ }$
		7	$\underline{\ \ } \bigcirc \underline{\ \ } = \underline{\ \ }$
3	$8 + 7 = \underline{\ \ }$	7	$\underline{\ \ } \bigcirc \underline{\ \ } = \underline{\ \ }$
		8	$\underline{\ \ } \bigcirc \underline{\ \ } = \underline{\ \ }$

4 ✎ Write **About It!** $9 + 8 = 17$. Write the related subtraction facts.

© Macmillan/McGraw-Hill

Practice

Use  and ▫. Add. Then subtract.
Write the related subtraction facts.

5

Add	Subtract	Related subtraction facts
6 + 7 = _13_	7	_13_ ⊝ _7_ = _6_
	6	_13_ ⊝ _6_ = _7_

---

6 8 + 6 = ____

____ ◯ ____ = ____

____ ◯ ____ = ____

7 5 + 9 = ____

____ ◯ ____ = ____

____ ◯ ____ = ____

8 8 + 3 = ____

____ ◯ ____ = ____

____ ◯ ____ = ____

9 4 + 9 = ____

____ ◯ ____ = ____

____ ◯ ____ = ____

✓ Spiral Review and Test Prep

Choose the best answer.

10 9 + 6 = ■ + 9

17   15   6   3

◯   ◯   ◯   ◯

11 15 < ■

16   14   7   5

◯   ◯   ◯   ◯

 Math at Home: Your child practiced related addition and subtraction facts.
Activity: Have your child tell how 9 + 8 = 17 is related to 17 − 9 = 8 and 17 − 8 = 9.

Name_____

Learn You can use a related addition fact to help you subtract.

Math Word

related facts

$16 - 7 = \boxed{9}$

$\boxed{9} + 7 = 16$

To find $16 - 7$, think:
$\blacksquare + 7 = 16$
$9 + 7 = 16$
So, $16 - 7 = 9$.

$16 - 7 = 9$ and $9 + 7 = 16$ are related facts.

Try It Find each missing number.

① $13 - 5 = \boxed{8}$

$\boxed{8} + 5 = 13$

② $15 - 8 = \boxed{}$

$\boxed{} + 8 = 15$

③ $15 - 9 = \boxed{}$

$\boxed{} + 9 = 15$

④ $17 - 8 = \boxed{}$

$\boxed{} + 8 = 17$

⑤
$$\begin{array}{r} 12 \\ -\ 7 \\ \hline \boxed{} \end{array} \qquad \begin{array}{r} \boxed{} \\ +\ 7 \\ \hline 12 \end{array}$$

⑥
$$\begin{array}{r} 11 \\ -\ 5 \\ \hline \boxed{} \end{array} \qquad \begin{array}{r} \boxed{} \\ +\ 5 \\ \hline 11 \end{array}$$

⑦
$$\begin{array}{r} 16 \\ -\ 8 \\ \hline \boxed{} \end{array} \qquad \begin{array}{r} \boxed{} \\ +\ 8 \\ \hline 16 \end{array}$$

⑧ Write **About It!** Why is each pair of facts in Exercises 1– 7 called related facts?

© Macmillan/McGraw-Hill

Practice Find each missing number.

Use related facts to help.

9 $13 - 6 = \boxed{7}$

$\boxed{7} + 6 = 13$

10 $14 - 8 = \boxed{}$

$\boxed{} + 8 = 14$

11
$$\begin{array}{r} 11 \\ -\ 6 \\ \hline \boxed{} \end{array} \qquad \begin{array}{r} \boxed{} \\ +\ 6 \\ \hline 11 \end{array}$$

12
$$\begin{array}{r} 19 \\ -\ 9 \\ \hline \boxed{} \end{array} \qquad \begin{array}{r} \boxed{} \\ +\ 9 \\ \hline 19 \end{array}$$

13
$$\begin{array}{r} 17 \\ -\ 9 \\ \hline \boxed{} \end{array} \qquad \begin{array}{r} \boxed{} \\ +\ 9 \\ \hline 17 \end{array}$$

14
$$\begin{array}{r} 12 \\ -\ 5 \\ \hline \boxed{} \end{array} \qquad \begin{array}{r} \boxed{} \\ +\ 5 \\ \hline 12 \end{array}$$

15
$$\begin{array}{r} 18 \\ -\ 9 \\ \hline \boxed{} \end{array} \qquad \begin{array}{r} \boxed{} \\ +\ 9 \\ \hline 18 \end{array}$$

16
$$\begin{array}{r} 20 \\ -10 \\ \hline \boxed{} \end{array} \qquad \begin{array}{r} \boxed{} \\ +10 \\ \hline 20 \end{array}$$

Problem Solving **Critical Thinking**

Show Your Work

THINK SOLVE EXPLAIN

Draw a picture to solve.

17 Bob has 3 coins.
They are worth 40¢.
One coin is a .
What are the other coins?

_____

Math at Home: Your child learned that addition and subtraction are related.
Activity: Have your child count 16 beads. Hide 7 in your hand. Have your child count the number of beads that are left and write ☐ + 9 = 16. Have your child tell you the number of hidden beads. Repeat for other numbers.

Learn A set of related facts is a fact family.

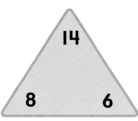

$$8 + 6 = \underline{14} \qquad 14 - 6 = \underline{8}$$

$$6 + 8 = \underline{14} \qquad 14 - 8 = \underline{6}$$

6, 8, and 14 make up this fact family.

Try It Add or subtract. Complete each fact family.

1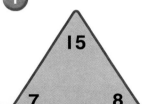

$$8 + 7 = \underline{15}$$

$$7 + 8 = \underline{15}$$

$$15 - 7 = \underline{8}$$

$$15 - 8 = \underline{7}$$

2

$$9 + 4 = \underline{\hspace{1cm}}$$

$$4 + 9 = \underline{\hspace{1cm}}$$

$$13 - 4 = \underline{\hspace{1cm}}$$

$$13 - 9 = \underline{\hspace{1cm}}$$

3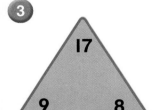

$$8 + 9 = \underline{\hspace{1cm}}$$

$$9 + 8 = \underline{\hspace{1cm}}$$

$$17 - 9 = \underline{\hspace{1cm}}$$

$$17 - 8 = \underline{\hspace{1cm}}$$

4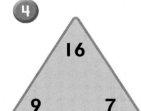

$$7 + 9 = \underline{\hspace{1cm}}$$

$$9 + 7 = \underline{\hspace{1cm}}$$

$$16 - 9 = \underline{\hspace{1cm}}$$

$$16 - 7 = \underline{\hspace{1cm}}$$

5 Write **About It!** What fact family can you make with the numbers 9, 5, and 14?

© Macmillan/McGraw-Hill

Practice Add or subtract.
Complete each fact family.

> Fact families use the same numbers.

6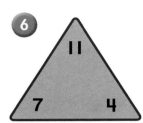
11 / 7 / 4

$7 + 4 = \underline{11}$

$11 - 4 = \underline{7}$

_____ + _____ = _____

_____ − _____ = _____

7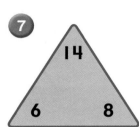
14 / 6 / 8

$6 + 8 = \underline{}$

$14 - 8 = \underline{}$

_____ + _____ = _____

_____ − _____ = _____

8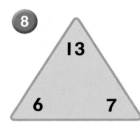
13 / 6 / 7

$6 + 7 = \underline{}$

_____ − _____ = _____

_____ + _____ = _____

_____ − _____ = _____

9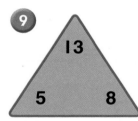
13 / 5 / 8

_____ + _____ = _____

_____ − _____ = _____

_____ + _____ = _____

_____ − _____ = _____

Problem Solving Critical Thinking

 THINK SOLVE EXPLAIN

10 Complete the fact family.
Use the numbers 9, 6, and 15.

_____ ◯ _____ = _____

_____ ◯ _____ = _____

_____ ◯ _____ = _____

_____ ◯ _____ = _____

Math at Home: Your child learned about fact families.
Activity: Have your child draw a row of 8 red flowers and a row of 2 blue flowers. Then have him or her write 2 addition facts and 2 subtraction facts using the numbers 8, 2, and 10.

Name_____

Find the shapes. Color.

circle rectangle square triangle trapezoid

Extra Practice

Use an inch ruler to measure.

about

_____ inches

about _____ inch

about _____ inches

about _____ inches

about _____ inches

LOG ON **www.mmhmath.com**
For more practice

Math at Home: Your child used a ruler to measure.
Activity: Have your child measure three objects. Then have your child put the objects in size order.

502   five hundred two

Name_____

Addition and Subtraction Patterns

Learn You can use patterns to add or subtract.

If you know 7 + 4 = 11, then you can add 17 + 4, 27 + 4, and 37 + 4.

If you know 7 − 4 = 3, then you can subtract 17 − 4, 27 − 4, and 37 − 4.

7	17	27	37
+4	+ 4	+ 4	+ 4
11	21	31	41

There is a pattern in the sums.

7	17	27	37
−4	− 4	− 4	− 4
3	13	23	33

There is a pattern in the differences.

Try It Use a pattern to add or subtract.

1.
8	18	28	38	48	58
+9	+ 9	+ 9	+ 9	+ 9	+ 9
17					

2.
6	16	26	36	46	56
−2	− 2	− 2	− 2	− 2	− 2
4					

3. **Write About It!** Look at row 2. What comes next in the pattern?

④
8	18	28	38
−3	− 3	− 3	− 3
5	15	25	35

> If you know 8 − 3 = 5, then you can use a pattern to find 18 − 3, 28 − 3, and 38 − 3.

⑤
7	17	27	37	47	57
+3	+ 3	+ 3	+ 3	+ 3	+ 3

⑥
5	15	25	35	45	55
−4	− 4	− 4	− 4	− 4	− 4

⑦
6	16	26	36	46	56
−3	− 3	− 3	− 3	− 3	− 3

Problem Solving ⟨ Estimation

⑧ About how many  are there?
Circle your estimate.

about 20   about 30

Math at Home: Your child used number patterns to add and subtract.
Activity: Ask your child: What is 3 + 5? Then ask: What is 13 + 5? Have your child tell you the number pattern.

Our Garden

There was an empty lot on our block. My friends and I made a garden there. Jon planted 9 tomato and 8 lettuce plants.

Problem Solving

 Problem and Solution

1. What did they do to the empty lot?

_____

2. Write a number sentence to find how many vegetables Jon planted in all? _____ + _____ = _____ vegetables

3. Mike planted 19 carrot and 8 pepper plants. How many did he plant in all? _____ plants

Garden Jobs

Growing a garden is a lot of work.
We share the work to get it done.
3 children water and 2 plant.

Problem and Solution

5. How do the children make their garden work?

_____

6. How many children are working
   in the garden?   _____ children

7. 27 flowers grow in the garden.
   If you added 10 more flowers, how many
   flowers would be in the garden? 27 + 10 = _____ flowers

Math at Home: Your child solved problems by finding a solution.
Activity: Have your child use the illustration to make up and solve another addition problem.

Solve.

1 Tim plants 8 ✿.
Pat plants 9 ✿.
How many ✿ do they
plant in all?

_____ + _____ = _____ ✿

2 Ann picks 8 🍎.
Her mom picks 19 🍎.
How many 🍎 do they
pick in all?

_____ + _____ = _____ 🍎

3 James plants 6 carrot
seeds.
Jill plants 8 carrot seeds.
How many carrot seeds do
they plant in all?

_____ + _____ = _____ seeds

4 Min pulls 6 weeds from
the garden.
Her dad pulls 18 weeds.
How many weeds do they
pull in all?

_____ + _____ = _____ weeds

Problem Solving

THINK
SOLVE
EXPLAIN

Write a Story!

5 Add. Tell about the pattern you see.

10 + 8 = _____   10 + 18 = _____   10 + 28 = _____

_____

_____

_____

Writing for Math

THINK
SOLVE
EXPLAIN

Look at the picture. Write an addition or subtraction problem about the picture.

Writing

Think

What addition or subtraction fact do I see in the picture?

_____ ◯ _____ ◯ _____

Solve

I can write my addition or subtraction problem now.

_____

_____

_____

Explain

I can tell you how my problem matches my fact.

_____

_____

Name_____

Add or subtract. You can use 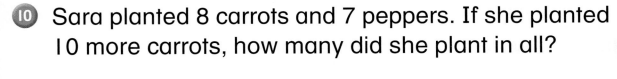.

1
```
  10      16
+  6    –  6
```

2
```
  8      11
+3     –  3
```

3
```
  9      14
+5     –  5
```

4
```
  6      12
+6     –  6
```

5
```
  8      15
+7     –  8
```

6
```
  6       9
+3      –6
```

Find the missing number.

7
```
  12    ☐
–  5   + 5
 ☐      12
```

8
```
  18    ☐
–  9   + 9
 ☐      18
```

9
```
  20    ☐
– 10  + 10
 ☐      20
```

Assessment

Solve.

10 Sara planted 8 carrots and 7 peppers. If she planted 10 more carrots, how many did she plant in all?

_____ plants

Spiral Review and Test Prep

Chapters 1–27

Choose the best answer.

1 Which number sentence belongs in this fact family?

| $5 + 6 = 11$ | $6 + 5 = 11$ | $11 - 6 = 5$ |

$11 + 6 = 17$ $5 + 11 = 16$ $11 - 5 = 6$

⬭   ⬭   ⬭

---

2 Katie wants to measure how long the table is.
Which tool should she use?

⬭   ⬭   ⬭

Complete the sentence.

3 A square has

_____ equal sides.

4 What time is it?

_____ : _____

THINK SOLVE EXPLAIN

5 Julie has a bag of and .
She says it is impossible to pick a .
Is Julie right? Why?
Use pictures or words to explain your answer.

_____

_____

Exploring Two-Digit Addition and Subtraction

Garden Song

Sung to the tune of "Skip to My Lou"

I have a garden where I can go—

Daisies and daffodils, row after row.

Lilies and tulips—but all of them grow

Faster than I can count them!

I'm counting the lily buds, iris and rose—

30 of this kind, and 20 of those.

As soon as I add them, another one grows,

Faster than I can count them!

Math at Home

Dear Family,

I will learn ways of adding and subtracting 2-digit numbers in Chapter 28. Here are my math words and an activity that we can do together.

Love, _____

My Math Words

count on :

Add 3 + 14.

Start with the greater number.

Count on 3.
Say 15, 16, 17.

3 + 14 = 17

count back :

Subtract 16 − 2.

Count back 2.

Say 15, 14.

16 − 2 = 14

Home Activity

Play a game with dimes and pennies. Your child adds 1, 2, or 3 pennies to 1, 2, or 3 dimes. Have your child count on from the greater number. Each time have him or her add another dime or penny.

Books to Read

In addition to these libary books, look for the Time for Kids math story that your child will bring home at the end of this unit.

- **Fifty on the Zebra** by Nancy Maria Grande Tabor, Charlesbridge, 1994.
- **Mrs. McTats and Her Houseful of Cats** by Alyssa Satin Capucilli, Margaret McElderry Books, 2001.
- **Time for Kids**

TIME FOR KIDS

Picnic Fun

Let's go on a picnic. We can bring food to share.

LOG ON
www.mmhmath.com
For Real World Math Activities

Add and Subtract Tens

HANDS ON Activity

Learn You can use to add or subtract tens.

Add 30 + 20.

Count on to add tens. Start at 30. Count on 2 tens and say 40, 50. So, 30 + 20 = 50.

3 tens + 2 tens = __5__ tens

30 + 20 = __50__

Subtract 50 − 10.

Count back to subtract tens. Start at 50. Count back 1 ten and say 40. So, 50 − 10 = 40.

5 tens − 1 ten = __4__ tens

50 − 10 = __40__

Your Turn Use . Count on to add tens. Count back to subtract tens.

1. 7 tens + 2 tens = __9__ tens

   70 + 20 = ___

2. 5 tens + 1 ten = ___ tens

   50 + 10 = ___

3. 9 tens − 2 tens = ___ tens

   90 − 20 = ___

4. 6 tens − 1 ten = ___ tens

   60 − 10 = ___

5. Write **About It!** How can you use to show 30 + 10?

6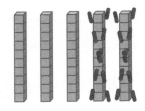

Count back to subtract tens.
Start at 50. Count back 2 tens
and say 40, 30.

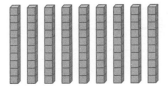

$50 - 20 =$ 30

---

7 $70 + 10 =$ _____

8 $20 + 10 =$ _____

9 $40 + 20 =$ _____

10 $30 + 30 =$ _____

11 $30 - 10 =$ _____

12 $70 - 20 =$ _____

13 $60 - 30 =$ _____

14 $90 - 10 =$ _____

Problem Solving **Reasoning** **Show Your Work**

15 Draw what was added.
Complete the number sentence.

_____ + _____ = _____

 Math at Home: Your child added and subtracted multiples of 10.
Activity: Say the number *70.* Ask your child to count on 2 tens. Then ask your child to count back 2 tens.

Name_____

Learn   You can use a hundred chart to add.

Math Word
_____
count on

1	2	3	4	5	6	7	8	9	10
11	12	13	14	15	16	17	18	19	20
21	22	23	24	25	26	27	28	29	30
31	32	33	34	35	36	37	38	39	40
41	42	43	44	45	46	47	48	49	50
51	52	53	54	55	56	57	58	59	60
61	62	63	64	65	66	67	68	69	70
71	72	73	74	75	76	77	78	79	80
81	82	83	84	85	86	87	88	89	90
91	92	93	94	95	96	97	98	99	100

$16 + 3 = \underline{19}$

Start at 16.
Count on 3
ones. Say
17, 18, 19.

$58 + 3 = \underline{61}$

Start at 58.
Count on 3 ones.
Say 59, 60.
Move to the next
row. Say 61.

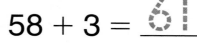

Try It   Count on to add. Use the hundred chart.

1. $44 + 2 = \underline{46}$

2. $14 + 3 = \underline{}$

3. $62 + 3 = \underline{}$

4. $18 + 3 = \underline{}$

5. $30 + 1 = \underline{}$

6. $79 + 2 = \underline{}$

7.   Write **About It!** How would you add $54 + 3$
on the hundred chart?

Practice   Count on to add. Use the hundred chart.

$35 + 3 = \underline{38}$

Start at 35.
Count on 3 ones.
Say 36, 37, 38.

1	2	3	4	5	6	7	8	9	10
11	12	13	14	15	16	17	18	19	20
21	22	23	24	25	26	27	28	29	30
31	32	33	34	35	36	37	38	39	40
41	42	43	44	45	46	47	48	49	50
51	52	53	54	55	56	57	58	59	60
61	62	63	64	65	66	67	68	69	70
71	72	73	74	75	76	77	78	79	80
81	82	83	84	85	86	87	88	89	90
91	92	93	94	95	96	97	98	99	100

8 $55 + 3 = \underline{}$

9 $97 + 3 = \underline{}$

10 $66 + 2 = \underline{}$

11 $49 + 2 = \underline{}$

12 $73 + 2 = \underline{}$

13 $18 + 3 = \underline{}$

✓ Spiral Review and Test Prep

Choose the best answer.

14 What time is shown?

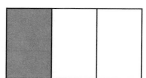

- ⬭ 1 o'clock
- ⬭ 5 o'clock
- ⬭ 6 o'clock
- ⬭ 12 o'clock

15 Which part is shaded?

▨		

$\dfrac{1}{4}$ ⬭ $\dfrac{1}{3}$ ⬭ $\dfrac{2}{3}$ ⬭ $\dfrac{3}{3}$ ⬭

Math at Home: Your child has been using a hundred chart to add.
Activity: Ask your child to explain how to add 43 + 3 on the hundred chart.

Name_____

Learn You can use a hundred chart to add.

1	2	3	4	5	6	7	8	9	10
11	12	13	14	15	16	17	18	19	20
21	22	23	24	25	26	27	28	29	30
31	32	33	34	35	36	37	38	39	40
41	42	43	44	45	46	47	48	49	50
51	52	53	54	55	56	57	58	59	60
61	62	63	64	65	66	67	68	69	70
71	72	73	74	75	76	77	78	79	80
81	82	83	84	85	86	87	88	89	90
91	92	93	94	95	96	97	98	99	100

$10 + 20 = \underline{30}$

Start at 10.
Count on 2 tens.
Say 20, 30.

$27 + 30 = \underline{57}$

Start at 27.
Count on 3 tens.
Say 37, 47, 57.

Try It Count on to add. Use the hundred chart.

1. $70 + 20 = \underline{90}$

2. $84 + 10 = \underline{}$

3. $45 + 30 = \underline{}$

4. $13 + 30 = \underline{}$

5. $50 + 20 = \underline{}$

6. $28 + 10 = \underline{}$

7. Write **About It!** How can you use the hundred chart to add 69 + 30?

Count on to add. Use the hundred chart.

$25 + 30 = \underline{55}$

Add 25 + 30.
Start at 25.
Count on 3 tens.
Say 35, 45, 55.

1	2	3	4	5	6	7	8	9	10
11	12	13	14	15	16	17	18	19	20
21	22	23	24	25	26	27	28	29	30
31	32	33	34	35	36	37	38	39	40
41	42	43	44	45	46	47	48	49	50
51	52	53	54	55	56	57	58	59	60
61	62	63	64	65	66	67	68	69	70
71	72	73	74	75	76	77	78	79	80
81	82	83	84	85	86	87	88	89	90
91	92	93	94	95	96	97	98	99	100

8. $44 + 10 = \underline{\hphantom{00}}$

9. $73 + 20 = \underline{\hphantom{00}}$

10. $18 + 30 = \underline{\hphantom{00}}$

11. $60 + 10 = \underline{\hphantom{00}}$

12. $15 + 30 = \underline{\hphantom{00}}$

13. $55 + 20 = \underline{\hphantom{00}}$

Problem Solving Number Sense

14. Continue the pattern.
Count by tens. You can use the hundred chart.

12, 22, 32, 42, _____, _____, _____

5, 15, 25, _____, _____, _____, _____

Math at Home: Your child has been using a hundred chart to add.
Activity: Ask your child to explain how to add 62 + 20 on the hundred chart.

Name_____

Learn You can use a hundred chart to subtract.

Math Word
_____
count back

1	2	3	4	5	6	7	8	9	10
11	12	13	14	15	16	17	18	19	20
21	22	23	24	25	26	27	28	29	30
31	32	33	34	35	36	37	38	39	40
41	42	43	44	45	46	47	48	49	50
51	52	53	54	55	56	57	58	59	60
61	62	63	64	65	66	67	68	69	70
71	72	73	74	75	76	77	78	79	80
81	82	83	84	85	86	87	88	89	90
91	92	93	94	95	96	97	98	99	100

$26 - 3 = \underline{23}$

Start at 26.
Count back 3 ones.
Say 25, 24, 23.

$62 - 3 = \underline{59}$

Start at 62. Count back
3 ones. Say 61. Then
move to the row above
and say 60, 59.

Try It Count back to subtract. Use the hundred chart.

① $64 - 2 = \underline{62}$

② $34 - 3 = \underline{}$

③ $76 - 1 = \underline{}$

④ $82 - 3 = \underline{}$

⑤ $95 - 2 = \underline{}$

⑥ $39 - 2 = \underline{}$

⑦ **Write About It!** How would you subtract $83 - 3$
on the hundred chart?

Practice Count back to subtract. Use the hundred chart.

$48 - 2 = \underline{46}$

Subtract 48 − 2. Start at 48. Count back 2 ones. Say 47, 46.

1	2	3	4	5	6	7	8	9	10
11	12	13	14	15	16	17	18	19	20
21	22	23	24	25	26	27	28	29	30
31	32	33	34	35	36	37	38	39	40
41	42	43	44	45	46	47	48	49	50
51	52	53	54	55	56	57	58	59	60
61	62	63	64	65	66	67	68	69	70
71	72	73	74	75	76	77	78	79	80
81	82	83	84	85	86	87	88	89	90
91	92	93	94	95	96	97	98	99	100

8 $65 - 3 = \underline{\hspace{1cm}}$

9 $55 - 3 = \underline{\hspace{1cm}}$

10 $79 - 1 = \underline{\hspace{1cm}}$

11 $60 - 1 = \underline{\hspace{1cm}}$

12 $36 - 2 = \underline{\hspace{1cm}}$

13 $72 - 3 = \underline{\hspace{1cm}}$

Make it Right

14 Here is how Julie subtracted. Tell what she did wrong. Make it right.

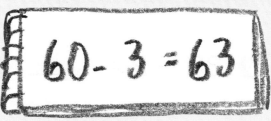

$60 - 3 = 63$

Math at Home: Your child has been using a hundred chart to subtract.
Activity: Ask your child to explain how to subtract 43 − 3 on the hundred chart.

520 five hundred twenty

Name _____

Learn You can use a hundred chart to subtract tens.

1	2	3	4	5	6	7	8	9	10
11	12	13	14	15	16	17	18	19	20
21	22	23	24	25	26	27	28	29	30
31	32	33	34	35	36	37	38	39	40
41	42	43	44	45	46	47	48	49	50
51	52	53	54	55	56	57	58	59	60
61	62	63	64	65	66	67	68	69	70
71	72	73	74	75	76	77	78	79	80
81	82	83	84	85	86	87	88	89	90
91	92	93	94	95	96	97	98	99	100

$50 - 30 = \underline{20}$

Start at 50. Count back 3 tens. Say 40, 30, 20.

$84 - 20 = \underline{64}$

Start at 84. Count back 2 tens. Say 74, 64.

Try It Count back to subtract. Use the hundred chart.

1. $90 - 20 = \underline{70}$

2. $68 - 30 = \underline{}$

3. $17 - 10 = \underline{}$

4. $53 - 20 = \underline{}$

5. $45 - 20 = \underline{}$

6. $26 - 10 = \underline{}$

7. ✎ Write **About It!** How can you use the hundred chart to subtract $79 - 30$?

Count back to subtract. Use the hundred chart.

$$31 - 30 = \underline{}$$

Subtract 31 − 30.
Start at 31.
Count back 3 tens.
Say 21, 11, 1.

1	2	3	4	5	6	7	8	9	10
11	12	13	14	15	16	17	18	19	20
21	22	23	24	25	26	27	28	29	30
31	32	33	34	35	36	37	38	39	40
41	42	43	44	45	46	47	48	49	50
51	52	53	54	55	56	57	58	59	60
61	62	63	64	65	66	67	68	69	70
71	72	73	74	75	76	77	78	79	80
81	82	83	84	85	86	87	88	89	90
91	92	93	94	95	96	97	98	99	100

8 $15 - 10 = \underline{5}$

9 $80 - 30 = \underline{}$

10 $99 - 20 = \underline{}$

11 $35 - 10 = \underline{}$

12 $52 - 30 = \underline{}$

13 $74 - 20 = \underline{}$

Problem Solving ⟨ Number Sense

14 Continue the patterns.
Count back by tens. You can use the hundred chart.

70, 60, 50, 40, _____, _____, _____

99, 89, 79, _____, _____, _____, _____

Math at Home: Your child has been using a hundred chart to subtract.
Activity: Ask your child to explain how to subtract 75 − 30.

Name_____

 **Learn** You can estimate sums and differences.

Will 10 + 12 be
greater than 20?

Will 20 − 3 be
greater than 20?

Think: 10 + 10 = 20.
Since 12 is greater than
10, the sum of 10 + 12
is greater than 20.

Think: When you subtract
a number from 20 the
difference is less than 20.

 **Try It** Solve. Circle yes or no.

1 Will 20 + 25 be greater than 40?   **Yes.**   No.

_____

2 Will 30 + 32 be less than 60?   Yes.   No.

_____

3 Will 50 − 10 be greater than 50?   Yes.   No.

_____

4 Will 75 − 20 be less than 70?   Yes.   No.

5 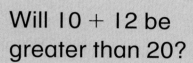 Write **About It!** Will 40 + 43 be greater than 80? Explain.

Practice Solve. Circle yes or no.

Think first! 90 − 12 will be less than 90.

Think first! 36 + 30 will be greater than 36.

6 Will 100 − 5 be less than 100?   Yes.   No.

7 Will 50 + 10 be greater than 55?   Yes.   No.

8 Will 10 + 9 be greater than 20?   Yes.   No.

9 Will 40 − 20 be less than 30?   Yes.   No.

10 Will 60 + 10 be greater than 70?   Yes.   No.

Problem Solving   Estimation

11 Will 40 − 20 be less than 40?
Explain.

_____

_____

Math at Home: Your child estimated sums and differences.
Activity: Ask your child, "Will 50 + 13 be greater than 60?" Have your child explain.

Name_____

Guess and Check

You can guess and check to help you solve problems.

Tara plants two packets of seeds. She plants a total of 23 seeds. Which two packets of seeds does she plant?

Read

What do I already know?   Tara planted _____ seeds.

What do I need to find?   _____

_____

Plan

I can guess and check.

Solve

I can carry out my plan.

$13 + 3 = 16$ No.
$3 + 10 = 13$ No.
$13 + 10 = 23$ Yes.

Look Back

Does my answer make sense?   Yes.   No.

How do I know? _____

© Macmillan/McGraw-Hill

Guess and check to solve.
Circle your answer.

1 Evan plants two packets of seeds. He plants a total of 55 seeds. Which two packets does he plant?

LETTUCE 25   ROSEMARY 30

BASIL 10

Problem Solving

2 Tom sees two kinds of birds in the garden. He sees 42 birds in all. Which two birds does he see?

39   2   40

3 Lawrence sees two kinds of bugs in the garden. He sees 66 bugs in all. Which two bugs does he see?

30   36   40

4 Amy picks pink and orange flowers in the garden. She picks 38 flowers in all. Which two flowers does she pick?

18   30   20

Math at Home: Your child solved problems by using guess and check.
Activity: Show your child 40 beans. Have your child guess which two numbers could show the total number of beans. Have your child check his or her answer.

Game Zone

Name_____

Hop Through the Garden

👥 2 players

How to Play:

▶ Pick a color counter. Put it on Start.

▶ Take turns. Flip a coin.

▶ Heads move 1 space. Tails move 2 spaces.

▶ Add or subtract. Your partner checks the answer.

▶ Correct answers score 1 point. The player with the most points wins.

You Will Need

63 − 2 = ___

78 + 2 = ___

76 − 20 = ___

Start

99 − 30 = ___

18 + 3 = ___

80 − 30 = ___

46 − 3 = ___

Finish

52 − 20 = ___

40 + 30 = ___

84 + 3 = ___

65 + 30 = ___

Technology Link

Compare Data • Calculator

You Will Use

You can use a  to compare data.

Robin grew three kinds of vegetables in her garden.

tomatoes	carrots	peppers
18	24	11

1. How many more carrots than peppers did Robin grow?

Press

Use the data on the chart to answer the questions.

You can use a .

2. How many more tomatoes than peppers did Robin grow? _____

---

3. If Robin grew 10 fewer tomatoes, how many tomatoes would there be? _____

---

4. If Robin grew 30 more carrots, how many carrots would she have in all? _____

Name_____

Use ▭▭▭ to add or subtract tens.

1 9 tens − 1 ten = _____ tens

90 − 10 = _____

2 5 tens + 3 tens = _____ tens

50 + 30 = _____

Add or subtract. Use ▭▭▭ and ▫ or a hundred chart.

3 40 + 20 = _____

4 16 + 30 = _____

5 23 + 3 = _____

6 44 − 10 = _____

7 61 − 2 = _____

8 74 − 30 = _____

9 87 − 2 = _____

1	2	3	4	5	6	7	8	9	10
11	12	13	14	15	16	17	18	19	20
21	22	23	24	25	26	27	28	29	30
31	32	33	34	35	36	37	38	39	40
41	42	43	44	45	46	47	48	49	50
51	52	53	54	55	56	57	58	59	60
61	62	63	64	65	66	67	68	69	70
71	72	73	74	75	76	77	78	79	80
81	82	83	84	85	86	87	88	89	90
91	92	93	94	95	96	97	98	99	100

Assessment

Guess and check. Circle your answer.

10 Spencer sees two kinds of trees in the yard. He sees 26 trees in all. Which two trees does he see?

© Macmillan/McGraw-Hill

Spiral Review and Test Prep
Chapters 1–28

Choose the best answer.

1 Spencer had 19 marbles. He lost 3.
How many does Spencer have now?

39 ⭕   22 ⭕   16 ⭕

2 Maya has 1 dime. She wants to buy gum for 20 cents.
How much more money does Maya need?

5 cents ⭕   10 cents ⭕   20 cents ⭕

Add.

3 $8 + 8 =$ _____

Write the missing number.

4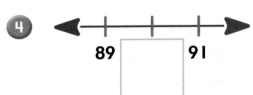

89 ☐ 91

THINK
SOLVE
EXPLAIN
Use pictures or words to explain your answer.

5 How can knowing $9 + 9 = 18$ help you know $18 - 9$?

_____

_____

Name _____

One fourth of the friends
play ball after lunch.
Color how many friends
are playing.

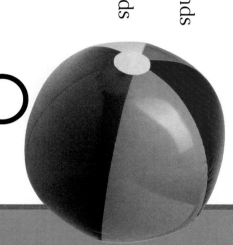

Picnic Fun

Let's go on a picnic.
We can bring food to share.

READ TOGETHER

© Macmillan/McGraw-Hill

Two friends share a watermelon slice.
Each friend gets one half.

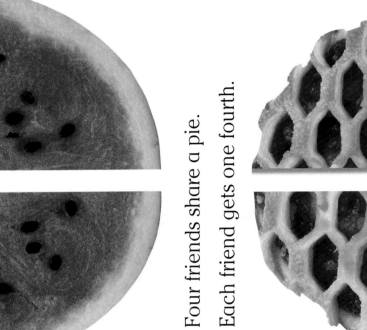

Four friends share a pie.
Each friend gets one fourth.

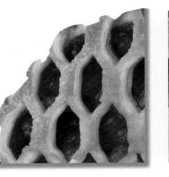

Three friends share a sandwich. They cut it into three equal parts. Each friend gets one third.

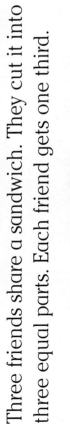

Name_____

Comparing and Temperature

Weather is what the air is like outside.
Weather can change.

Problem Solving

In the morning, the temperature may be cool.

The temperature rises during the day. The heat from the sun makes this happen.

Circle the word to complete each sentence.

1. _____ is how warm or cool something is.

   Weather   Temperature

2. _____ is what the air is like outside.

   Weather   Temperature

What to Do

- Go outside in the morning.

- Use a . Find the temperature. Record it.

- Go outside in the afternoon.

- Use a . Find the temperature. Record it.

Temperature	
Morning	_____ degrees
Afternoon	_____ degrees

Problem Solving

Solve.

③ **Compare** What happened to the temperature in the afternoon? Circle the answer.

It got warmer.   It got cooler.

④ **Compare** Which temperature is greater?

_____ degrees > _____ degrees

⑤ **Predict** What do you think would happen if you took the temperatures again tomorrow? Why?

_____

_____

_____

_____

Math at Home: Your child compared daily temperature changes.
Activity: Repeat this activity noting temperature differences in various places in your home throughout the day.

Name_____

Math Words

Draw lines to match.

1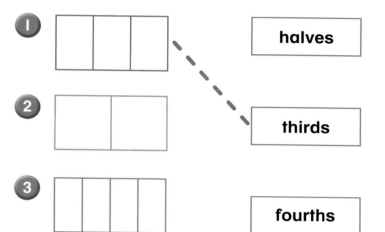

halves

2

thirds

3

fourths

Skills and Applications

Fractions (pages 453–464, 473–476)

Examples

Shade to show the fraction.

$\frac{1}{4}$ $\frac{2}{4}$ $\frac{3}{4}$

The top number tells how many shaded parts. The bottom number tells how many parts in all.

4

$\frac{1}{5}$

5

$\frac{3}{6}$

Which is greater?

$\frac{1}{4}$ $\frac{1}{2}$

$\frac{1}{2}$ is greater than $\frac{1}{4}$.

6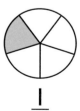

$\frac{1}{3}$ $\frac{1}{5}$

_____ is greater than _____

Skills and Applications

Addition and Subtraction (pages 491–504, 513–524)

Examples

Unit Review

Use fact families.

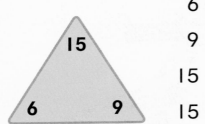

$6 + 9 = 15$

$9 + 6 = 15$

$15 - 9 = 6$

$15 - 6 = 9$

7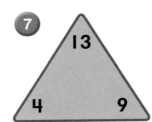

$4 + 9 = \underline{\hphantom{00}}$

$9 + 4 = \underline{\hphantom{00}}$

$13 - 9 = \underline{\hphantom{00}}$

$13 - 4 = \underline{\hphantom{00}}$

Use related facts.

$5 + 9 = 14$ $10 + 6 = 16$

$14 - 9 = 5$ $16 - 6 = 10$

8 $8 + 7 = \underline{\hphantom{00}}$

$15 - 7 = \underline{\hphantom{00}}$

(pages 483–484, 525–526)

Problem Solving   Strategy

Guess and check to solve.

| 10 bugs | 1 bug | 11 bugs |

Robin sees 21 bugs. Which two kinds of bugs does she see?

$10 + 1$ is not 21.

$11 + 1$ is not 21.

$11 + 10$ is 21.

9 Richard sees 43 birds. Circle the two kinds of birds he sees.

| 20 birds | 3 birds | 23 birds |

Math at Home: Your child reviewed fractions and addition and subtraction.
Activity: Have your child use these pages to review facts.

Pattern Block Fractions

Draw lines in each shape.
Color to show the fraction.

You Will Need

1 Use pattern blocks to show 2 equal parts. Draw a line to show 2 equal parts. Color.

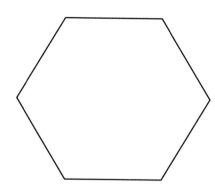

2 Use pattern blocks to show 3 equal parts. Draw lines to show 3 equal parts. Color.

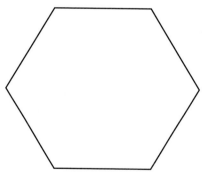

3 Use pattern blocks to show 6 equal parts. Draw lines to show 6 equal parts. Color.

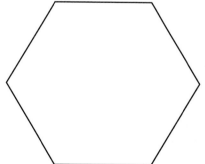

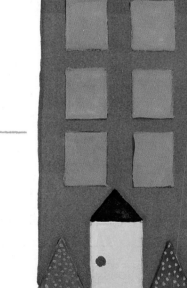

Portfolio You may want to put this page in your portfolio.

Assessment

Unit 7
Enrichment

Add 2-Digit Numbers

Add 25 + 3.

Add the ones first.
Then add the tens.

tens	ones
2	5
+	3
2	8

Add 26 + 22.

Add the ones first.
Then add the tens.

tens	ones
2	6
+ 2	2
4	8

Add. You can use 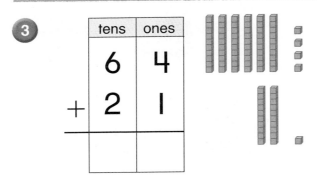 and ▫.

1

tens	ones
4	3
+	2

2

tens	ones
1	5
+	4

3

tens	ones
6	4
+ 2	1

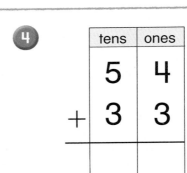

4

tens	ones
5	4
+ 3	3

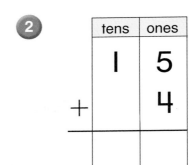

Picture Glossary

add (page 71)

● ●   ● ● ●

$$2 + 3 = 5$$

$$\begin{array}{r} 2 \\ + 3 \\ \hline 5 \end{array}$$

afternoon (page 331)

addend (page 179)

$$\begin{array}{r} 5 \leftarrow \text{addend} \\ + 3 \leftarrow \text{addend} \\ \hline 8 \end{array}$$

bar graph (page 195)

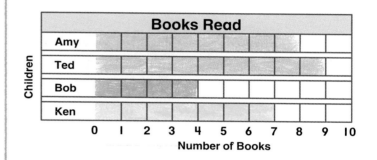

addition sentence (page 53)

$$5 + 3 = 8$$

before (page 37)

just before 4

after (page 37)

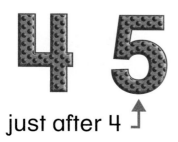

just after 4

between (page 37)

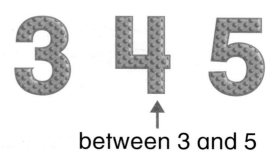

between 3 and 5

Glossary

Picture Glossary

calendar (page 351)

June						
S	M	T	W	T	F	S
			1	2	3	4
5	6	7	8	9	10	11
12	13	14	15	16	17	18
19	20	21	22	23	24	25
26	27	28	29	30		

circle (page 411)

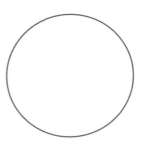

cent (¢) (page 255)

1 ¢   1 cent

closed shape (page 431)

starts and ends at the same point

centimeter (page 379)

1 centimeter

compare (page 33)

 < =  >

5 is less   6 is equal   8 is greater
than 7   to 6   than 4

certain (page 481)

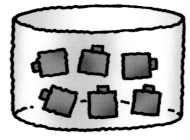

It is certain you will pick a .

cone (page 409)

Picture Glossary

count back (page 159)

$$9 - 2 = 7$$

count on (page 143)

$$7 + 2 = 9$$

cube (page 409)

cup (page 391)

I cup

curve (page 415)

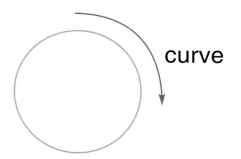

curve

cylinder (page 409)

data (page 5)

Favorite Colors	
Color	Number of Children
blue	⊮⊮
red	|||

Data is information.

days of the week (page 351)

May						
S	M	T	W	T	F	S
1	2	3	4	5	6	7
8	9	10	11	12	13	14
15	16	17	18	19	20	21
22	23	24	25	26	27	28
29	30	31				

days of the week

Glossary

Picture Glossary

degrees (page 399)

70 degrees Fahrenheit

difference (page 103)

$$16 - 9 = 7$$

$$\begin{array}{r} 16 \\ -\ 9 \\ \hline 7 \end{array}$$

↑ difference →

Subtract to find the difference.

dime (page 257)

10¢   10 cents

dollar (page 275)

dollar bill
100¢
$1.00

dollar coin
100¢
$1.00

doubles (page 179)

4 + 4 = 8

doubles plus 1 (page 299)

4 + 5 = 9

edge (page 409)

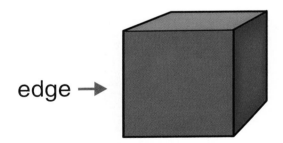

edge →

equal parts (page 453)

2 equal parts

Glossary

Picture Glossary

equally likely (page 477)

It is equally likely to spin ⬤ or ⬤.

evening (page 331)

equals (=) (page 53)

$$2 + 3 = 5$$

equals ↰

face (page 409)

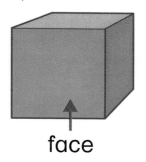

face

estimate (page 147)

There are about 10.

fact family (page 125)

$$6 + 7 = 13 \quad 13 - 7 = 6$$
$$7 + 6 = 13 \quad 13 - 6 = 7$$

even (page 245)

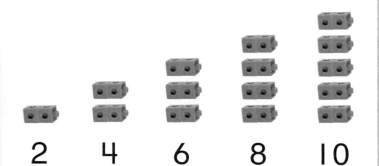

2   4   6   8   10

fewer (page 33)

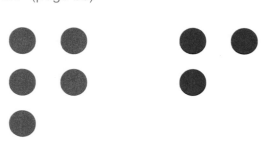

There are fewer ⬤.

Glossary

Picture Glossary

flip (page 435)

a mirror image of a figure

foot (page 375)

12 inches = 1 foot

fraction (page 455)

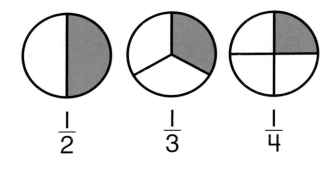

$\frac{1}{2}$ $\frac{1}{3}$ $\frac{1}{4}$

gram (page 397)

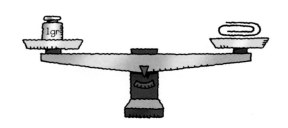

1 gram is about 1 paper clip.

half hour (page 337)

30 minutes

half past (page 337)

It is half past 7.

heavier (page 389)

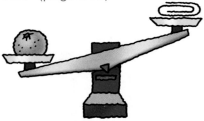

An orange is heavier
than a paper clip.

hour (page 339)

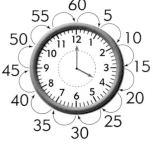

60 minutes

Picture Glossary

hour hand (page 333)

hour hand

impossible (page 481)

It is impossible to pick a .

inch (page 373)

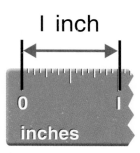

1 inch

0   1
inches

is equal to (=) (page 237)

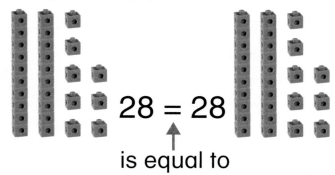

28 = 28

is equal to

is greater than (>) (page 35)

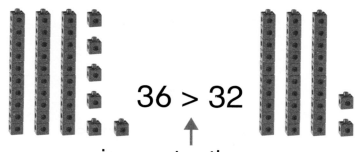

36 > 32

is greater than

is less than (<) (page 35)

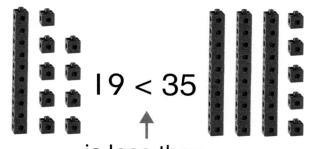

19 < 35

is less than

kilogram (page 397)

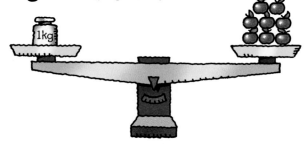

1 kilogram is about 8 apples.

less likely (page 479)

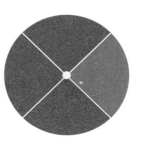

It is less likely to spin ●.

© Macmillan/McGraw-Hill

Glossary

Picture Glossary

lighter (page 389)

A paper clip is lighter than an orange.

line of symmetry (page 437)

A line of symmetry makes two matching parts.

liter (page 395)

1 liter

longest (page 371)

long

longer

longest

measure (page 371)

6 cubes long

minus (−) (page 55)

$$6 - 1 = 5$$

↑ minus

minute hand (page 333)

minute hand

minutes (page 339)

60 minutes equal 1 hour.

Glossary

Picture Glossary

mode (page 203)

the number that occurs most often in a set of data

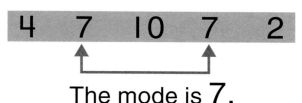

The mode is 7.

morning (page 331)

month (page 351)

	November					
S	M	T	W	T	F	S
		1	2	3	4	5
6	7	8	9	10	11	12
13	14	15	16	17	18	19
20	21	22	23	24	25	26
27	28	29	30			

This calendar shows the month of November.

nickel (page 255)

5¢   5 cents

more (page 33)

There are more ●.

number (page 17)

A number tells how many.

 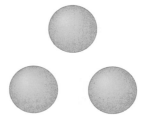

more likely (page 479)

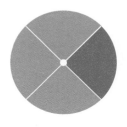

It is more likely you will spin ●.

number line (page 37)

Glossary

© Macmillan/McGraw-Hill

Picture Glossary

o'clock (page 333)

 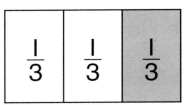

7 o'clock

one third (page 455)

$\frac{1}{3}$

One third is shaded.

odd (page 245)

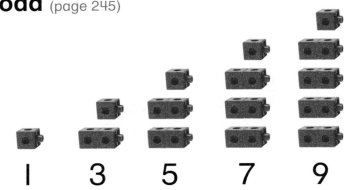

1   3   5   7   9

ones (page 21)

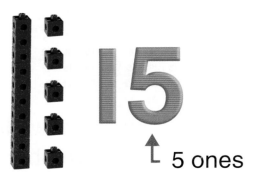

5 ones

one fourth (page 455)

$\frac{1}{4}$

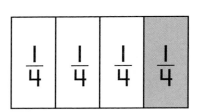

One fourth is shaded.

open shape (page 431)

starts and ends at different points

one half (page 455)

$\frac{1}{2}$

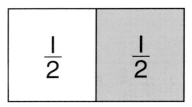

One half is shaded.

order (page 37)

1, 2, 3, 4, 5

These numbers are
in counting order.

Glossary

Picture Glossary

ordinal number (page 41)

first   second   third

1st   2nd   3rd

pint (page 391)

2 cups equal 1 **pint.**

pattern unit (page 439)

the repeating part of a pattern

plus (+) (page 53)

$$2 + 3 = 5$$

↑ plus

penny (page 255)

1¢   1 cent

pound (page 393)

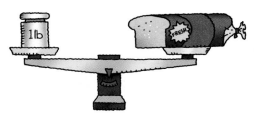

The bread weighs about 1 pound.

picture graph (page 7)

Class Pets					
dog					
cat					
bird					

predict (page 479)

telling what you think will happen

Glossary

© Macmillan/McGraw-Hill

Picture Glossary

probable (page 481)

It is probable you will pick a .

range (page 203)
the difference between the greatest and least number in a set of data

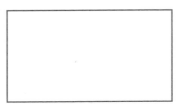

$$10 - 2 = 8$$

The range is 8.

pyramid (page 409)

rectangle (page 411)

quart (page 391)

4 cups equal 1 quart.

rectangular prism (page 409)

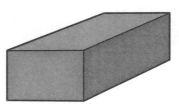

quarter (page 273)

25¢   25 cents

regroup (page 221)

10 ones = 1 ten